T4-ALD-413

Lecture Notes in Computer Science

Edited by G. Goos and J. Hartmanis

P-0129-24
6-2-86
IN

157

Ole Østerby
Zahari Zlatev

Direct Methods for Sparse Matrices

RETIRE DE LA COLLECTION - UQO

UNIVERSITÉ DU QUÉBEC À HULL BIBLIOTHÈQUE

Springer-Verlag
Berlin Heidelberg New York Tokyo 1983

Editorial Board

D. Barstow W. Brauer P. Brinch Hansen D. Gries D. Luckham
C. Moler A. Pnueli G. Seegmüller J. Stoer N. Wirth

Authors

Ole Østerby
Computer Science Department, Aarhus University
DK 8000 Aarhus, Denmark

Zahari Zlatev
Air Pollution Laboratory
Danish Agency of Environmental Protection
Risø National Laboratory
DK 4000 Roskilde, Denmark

QA
188
O77
1983

06-10203761

CR Subject Classifications (1982): G.1.3

ISBN 3-540-12676-7 Springer-Verlag Berlin Heidelberg New York Tokyo
ISBN 0-387-12676-7 Springer-Verlag New York Heidelberg Berlin Tokyo

This work is subject to copyright. All rights are reserved, whether the whole or part of the material is concerned, specifically those of translation, reprinting, re-use of illustrations, broadcasting, reproduction by photocopying machine or similar means, and storage in data banks. Under § 54 of the German Copyright Law where copies are made for other than private use, a fee is payable to "Verwertungsgesellschaft Wort", Munich.

© by Springer-Verlag Berlin Heidelberg 1983

Printed in Germany

Printing and binding: Beltz Offsetdruck, Hemsbach/Bergstr.
2145/3140-543210

Preface

The mathematical models of many practical problems lead to systems of linear algebraic equations where the coefficient matrix is large and sparse. Typical examples are the solutions of partial differential equations by finite difference or finite element methods but many other applications could be mentioned.

When there is a large proportion of zeros in the coefficient matrix then it is fairly obvious that we do not want to store all those zeros in the computer, but it might not be quite so obvious how to get around it. We shall first describe storage techniques which are convenient to use with direct solution methods, and we shall then show how a very efficient computational scheme can be based on Gaussian elimination and iterative refinement.

A serious problem in the storage and handling of sparse matrices is the appearance of fill-ins, i.e. new elements which are created in the process of generating zeros below the diagonal. Many of these new elements tend to be smaller than the original matrix elements, and if they are smaller than a certain quantity which we shall call the drop tolerance we simply ignore them. In this way we may preserve the sparsity quite well but we probably introduce rather large errors in the LU decomposition to the effect that the solution becomes unacceptable. In order to retrieve the accuracy we use iterative refinement and we show theoretically and with practical experiments that it is ideal for the purpose.

Altogether, the combination of Gaussian elimination, a large drop tolerance, and iterative refinement gives a very efficient and competitive computational scheme for sparse problems. For dense matrices iterative refinement will always require more storage and computation time, and the extra accuracy it yields may not be enough to justify it. For sparse problems, however, iterative refinement combined with a large drop tolerance will in most cases give very accurate results and reliable error estimates with less storage and computation time.

A short description of the Gaussian elimination process is given in chapter 1. Different storage algorithms for general sparse matrices are discussed in chapter 2. Chapter 3 is devoted to the use of pivotal strategies as a tool for keeping the balance between sparsity and accuracy. The possibility of using an iterative refinement process in connection with the Gaussian elimination is the topic of chapter 4.

In chapter 5 we introduce a general computational scheme which includes many well-known direct methods for linear equations and for overdetermined linear systems as special cases. We also demonstrate how the above techniques can be generalized to linear least squares problems. Thus, we show that the theory of most of the direct methods can be studied from a common point of view and that the algorithms described in the previous chapters are applicable not only in connection with Gaussian elimination but also for many other methods. A particular algorithm (the Gentleman - Givens orthogonalization) is discussed in detail in the second part of chapter 5 as an illustration of the above statements.

The algorithms described in chapters 2 - 4 have been implemented in a package for the solution of large and sparse systems of linear algebraic equations. This package, Y12M, is included in the standard library at RECKU (the Regional Computing Centre at the University of Copenhagen). The subroutines of package Y12M with full documentation and with many test-programs are available at the usual cost (for the magnetic tape, machine time, shipment, etc.). Requests should be addressed to J. Wasniewski, RECKU, Vermundsgade 5, DK - 2100 Copenhagen. It should be mentioned that the subroutines are written in FORTRAN. Both double and single precision versions are available. No special features of the computer at the disposal at RECKU (UNIVAC 1100/82) have been exploited and no machine-dependent constants are used. Thus the package is portable and will work without any changes on many large computers. This has been verified by running the subroutines of the package on three different computers: a UNIVAC 1100/82 computer at RECKU, an IBM 3033 computer at the Northern Europe University Computing Centre (NEUCC) and a CDC Cyber 173 computer at the Regional Computing Centre at Aarhus University (RECAU).

The package Y12M also includes subroutines for estimation of the condition number of a sparse matrix. The subroutines can be called when the LU decomposition is calculated and provide a relatively inexpensive but still reliable measure of the sensitivity of the results to round-off errors.

A full documentation of the subroutines from package Y12M with a brief description of the basic ideas applied in the implementation is given in a previous volume of this series (see Z. Zlatev, J. Wasniewski and K. Schaumburg: "Y12M - Solution of Large and Sparse Systems of Linear Algebraic Equations", Lecture Notes in Computer Science, Vol. 121, Springer, Berlin-Heidelberg-New York, 1981).

Decimal notation is used for the numbering of sections and chapters. Thus the third section of chapter 5 is numbered 5.3. The 15th numbered equation in section 3 of chapter 5 is numbered (3.15) and is referenced in another chapter by (5.3.15). Tables and figures are numbered in each chapter. Thus the 7th table or figure in chapter 1 is numbered 1.7. A similar numbering system is used for theorems, corollaries, remarks, etc.

We would like to express our thanks to Angelika Paysen who with great patience and expert skill typed the manuscript.

Contents

4. Iterative Refinement

5. Other Direct Methods

Appendix

Chapter 1: Introduction

1.1 Gaussian elimination

Many practical problems lead to large systems of linear algebraic equations

(1.1) $Ax = b,$

where $n \in \mathbb{N}$, $A \in \mathbb{R}^{nxn}$ and $b \in \mathbb{R}^{nx1}$ are given, with $\operatorname{rank}(A) = n$, and $x \in \mathbb{R}^{nx1}$ is to be computed.

In this book we shall discuss the solution of (1.1) by means of so-called direct methods and begin with the well known Gaussian elimination. The elimination process will be carried out in $n-1$ stages

(1.2) $$A^{(k+1)} = L^{(k)} \cdot A^{(k)}, \qquad (k = 1(1)n-1)$$

starting with $A^{(1)} = A$. The lower right $(n-k+1) \times (n-k+1)$ submatrix of $A^{(k)}$ is denoted A_k and its elements are denoted $a_{ij}^{(k)}$, $(i,j = k(1)n)$. For the elements of A_{k+1} we have the formula

(1.3) $$a_{ij}^{(k+1)} = a_{ij}^{(k)} - a_{ik}^{(k)} \cdot a_{kj}^{(k)} / a_{kk}^{(k)}, \qquad i,j = k+1(1)n.$$

$L^{(k)}$ is an elementary unit lower triangular matrix with elements

(1.4) $$\begin{aligned} l_{ii}^{(k)} &= 1, && (i = 1(1)n); \\ l_{ik}^{(k)} &= -a_{ik}^{(k)} / a_{kk}^{(k)}, && (i = k+1(1)n); \\ &\text{otherwise } 0. \end{aligned}$$

The end result of the elimination is the upper triangular matrix $U = A^{(n)}$ and the process is equivalent to a triangular factorization

(1.5) $$A = L \cdot U,$$

where

(1.6) $$L = (L^{(n-1)} \circ L^{(n-2)} \cdot \ldots \cdot L^{(1)})^{-1}.$$

The elements of L and U are thus given by

$$(1.7) \qquad U = \begin{bmatrix} a_{11}^{(1)} & a_{12}^{(1)} & a_{13}^{(1)} & & a_{1n}^{(1)} \\ & a_{22}^{(2)} & a_{23}^{(2)} & & a_{2n}^{(2)} \\ & & \cdot & & \cdot \\ & & & \cdot & \cdot \\ & & & & \cdot \\ & & & & a_{nn}^{(n)} \end{bmatrix},$$

and

$$(1.8) \qquad L = \begin{bmatrix} 1 & & & & \\ -l_{21}^{(1)} & 1 & & 0 & \\ -l_{31}^{(1)} & -l_{32}^{(2)} & 1 & & \\ \cdot & & \cdot & & \\ \cdot & & & \cdot & \\ \cdot & & & & \cdot \\ -l_{n1}^{(1)} & -l_{n2}^{(2)} & \cdots & -l_{n,n-1}^{(n-1)} & 1 \end{bmatrix}.$$

In order for this factorization to be successful it is necessary that all the denominators in (1.3), $a_{kk}^{(k)}$, be different from 0. Moreover, to ensure reasonably stable computations it is to be desired that the correction terms in (1.3), $a_{ik}^{(k)} \cdot a_{kj}^{(k)} / a_{kk}^{(k)}$ be reasonably small. This is usually accomplished by interchanging rows and/or columns and thus requiring that $|l_{ik}^{(k)}| \leq 1$ or $|a_{kj}^{(k)} / a_{kk}^{(k)}| \leq 1$. We shall return to this topic in section 3.1 and for the moment just prepare ourselves for the row and column interchanges which transform (1.1) into

$$(1.9) \qquad P A Q (Q^T x) = Pb,$$

where P and Q are permutation matrices.

The elimination or factorization (1.5) now becomes

$$LU = PAQ + E, \tag{1.10}$$

where L and U now denote the computed triangular matrices and E is a perturbation matrix which takes care of the computational errors, among other things.

An approximation $\tilde{x}$ to the solution x is now computed by substitution:

$$x_1 = QU^{-1}L^{-1}Pb, \tag{1.11}$$

and we set

$$\tilde{x} = x_1 \tag{1.12}$$

<u>Definition 1.1</u> $\tilde{x}$ as given by (1.12) is called the direct solution (DS). ∎

<u>Remark 1.2</u> Even if the computations in (1.11) are performed without errors we may still have $\tilde{x} \neq x$ if $E \neq 0$ in (1.10). ∎

We would expect that the process of elimination and substitution would lead to a 'good' solution if the elements of E are small. This is often the case but we have no a priori guarantee of this, and we don't even have any a priori guarantee that the elements of E will be small if we use only row-interchanges. Therefore the following 'refining' process can be useful.

Compute for $i = 1, 2, \ldots, q-1$

$$r_i = b - Ax_i, \tag{1.13}$$

$$d_i = QU^{-1}L^{-1}Pr_i, \tag{1.14}$$

$$x_{i+1} = x_i + d_i, \tag{1.15}$$

and set

$$\tilde{x} = x_q. \tag{1.16}$$

Definition 1.3 The process described by (1.13) - (1.15) is called iterative refinement. $\tilde{x}$ as given by (1.16) is called the iteratively refined solution (IR). ∎

Remark 1.4 Under certain conditions the process (1.13) - (1.15) is convergent and $x_i \to x (i \to \infty)$. In this case $x = x_1 + \Sigma_{i=1}^{\infty} d_i$ and $d_i \to 0$. If the series converges swiftly $\|d_i\|$ can be used as an estimate of the error $\|x - x_i\|$. ∎

If convergent the iterative refinement will provide a better solution and a reasonable error estimate. The price we have to pay for this is extra storage (because a copy of A must be retained) and extra computing time (for the process (1.13) - (1.15)). The following table gives the storage and computing time for DS and IR

	DS	IR
Storage	$n^2 + O(n)$	$2n^2 + O(n)$
Time	$\frac{1}{3}n^3 + n^2 + O(n)$	$\frac{1}{3}n^3 + (2q - 1)n^2 + O(n)$

Table 1.1

Comparison of storage and time with DS and IR for dense matrices. The computation time is measured by the number of multiplications.

1.2 Sparse matrices

Until now we have tacitly assumed that we require space and time to treat all the n^2 elements of matrix A (A is dense). Table 1.1 shows that in this case both storage and time increase rapidly with n and that IR is always more expensive than DS in both respects.

In many applications, however, A is sparse, i.e. a large proportion of the elements of A are 0, and we shall in this book describe special techniques which can be used to exploit this sparsity of A.

The border-line between dense and sparse matrices is rather fluent, but we could 'define' a matrix to be sparse if we can save space and/or time by employing the sparse matrix techniques to be described in this book.

Consider the basic formula in the factorization process (1.2)

(2.1) $$a_{ij}^{(k+1)} = a_{ij}^{(k)} - a_{ik}^{(k)} \cdot a_{kj}^{(k)} / a_{kk}^{(k)} \qquad (a_{kk}^{(k)} \neq 0)$$

$$i,j = k+1(1)n, \quad k = 1(1)n-1.$$

The computation is clearly simplified if one or more of the quantities involved (except $a_{kk}^{(k)}$) is 0.

A sparse matrix technique is based on the following main principles:

A) Only the non-zero elements of matrix A are stored.

B) We attempt to perform only those computations which lead to changes, i.e. we only use formula (2.1) when $a_{ik}^{(k)} \neq 0$ and $a_{kj}^{(k)} \neq 0$.

C) The number of 'new elements' (fill-ins) is kept small. A new element is generated when $a_{ij}^{(k)} = 0$ and $a_{ij}^{(k+1)} \neq 0$.

Before we continue we shall introduce some notation and terminology.

By an <u>element</u> of matrix A we mean a non-zero element of the matrix. The rest of matrix A are called zeros and are treated as such.

n denotes the number of unknowns (columns).

m denotes the number of equations (rows).
(We shall only treat the case $m \neq n$ in chapter 5.)

NZ denotes the number of elements of matrix A.

NN is the length of the one-dimensional array A which is used to hold the elements ($NN \geq NZ$).

COUNT is the maximum number of elements (including fill-ins) kept in array A during the elimination process ($NN \geq COUNT$).

T is the drop tolerance (see the end of section 1.4).

We shall see that the use of sparse matrix techniques will change the contents of table 1.1 completely. More specifically, the computation time and the storage will not grow as fast with n, the storage needed for IR will not always be larger than for DS (because we introduce the drop tolerance), and the computation time will often be smaller for IR than for DS with the techniques which we are going to describe in the following chapters.

1.3 Test matrices

More often than not assertions and suggestions about sparse matrix techniques cannot be proved mathematically. We shall often have to rely on practical experiments to show that one technique is better than another - or to see under which circumstances it is better. For this purpose several classes of test matrices have been constructed, either as typical examples or generalizations of practically occurring matrices, or as nasty examples designed to make life difficult for sparse matrix programs.

We shall in this section introduce some of those test matrices which we are going to use throughout the text.

Test matrices of class D(n,c) are $n \times n$ matrices with 1 in the diagonal, three bands at the distance c above the diagonal (and reappearing cyclicly under it), and a 10×10 triangle of elements in the upper right-hand corner.

More specifically:

$$a_{i,i} = 1, \qquad i = 1(1)n;$$

$$a_{i,i+c} = i+1, \qquad i = 1(1)n-c, \quad a_{i,i-n+c} = i+1, \quad i = n-c+1(1)n;$$

$$a_{i,i+c+1} = -i, \qquad i = 1(1)n-c-1, \quad a_{i,i-n+c+1} = -i, \quad i = n-c(1)n;$$

$$a_{i,i+c+2} = 16, \qquad i = 1(1)n-c-2, \quad a_{i,i-n+c+2} = 16 \quad i = n-c-1(1)n;$$

$$a_{i,n-11+i+j} = 100 \cdot j, \quad i = 1(1)11-j, \quad j = 1(1)10;$$

for any $n \geqq 14$ and $1 \leqq c \leqq n-13$.

By varying n and c we can obtain matrices of different sizes and sparsity patterns. In Fig. 1.2 we show the sparsity pattern of matrix D(20,5).

```
x o o o o x x x o o x x x x x x x x x x
o x o o o o x x x o o x x x x x x x x x
o o x o o o o x x x o o x x x x x x x x
o o o x o o o o x x x o o x x x x x x x
o o o o x o o o o x x x o o x x x x x x
o o o o o x o o o o x x x o o x x x x x
o o o o o o x o o o o x x x o o x x x x
o o o o o o o x o o o o x x x o o x x x
o o o o o o o o x o o o o x x x o o x x
o o o o o o o o o x o o o o x x x o o x
o o o o o o o o o o x o o o o x x x o o
o o o o o o o o o o o x o o o o x x x o
o o o o o o o o o o o o x o o o o x x x
x o o o o o o o o o o o o x o o o o x x
x x o o o o o o o o o o o o x o o o o x
x x x o o o o o o o o o o o o x o o o o
o x x x o o o o o o o o o o o o x o o o
o o x x x o o o o o o o o o o o o x o o
o o o x x x o o o o o o o o o o o o x o
o o o o x x x o o o o o o o o o o o o x
```

<u>Fig. 1.2</u>

Sparsity pattern of matrix D(20,5)

<u>Test matrices of class E(n,c)</u> are symmetric, positive definite, $n \times n$ matrices with 4 in the diagonal and -1 in the two sidediagonals and in two bands at the distance c from diagonal. These matrices are rather similar to matrices obtained from using the five-point formula in the discretization of elliptic partial differential equations.

$$
\begin{aligned}
&a_{ii} = 4, && i = 1(1)n; \\
(3.2)\qquad &a_{i,i+1} = a_{i+1,i} = -1, && i = 1(1)n-1; \\
&a_{i,i+c} = a_{i+c,i} = -1, && i = 1(1)n-c;
\end{aligned}
$$

where $n \geqq 3$ and $2 \leqq c \leqq n-1$.

In Fig. 1.3 we show the matrix E(10,4)

$$\begin{matrix} 4 & -1 & 0 & 0 & -1 & 0 & 0 & 0 & 0 & 0 \\ -1 & 4 & -1 & 0 & 0 & -1 & 0 & 0 & 0 & 0 \\ 0 & -1 & 4 & -1 & 0 & 0 & -1 & 0 & 0 & 0 \\ 0 & 0 & -1 & 4 & -1 & 0 & 0 & -1 & 0 & 0 \\ -1 & 0 & 0 & -1 & 4 & -1 & 0 & 0 & -1 & 0 \\ 0 & -1 & 0 & 0 & -1 & 4 & -1 & 0 & 0 & -1 \\ 0 & 0 & -1 & 0 & 0 & -1 & 4 & -1 & 0 & 0 \\ 0 & 0 & 0 & -1 & 0 & 0 & -1 & 4 & -1 & 0 \\ 0 & 0 & 0 & 0 & -1 & 0 & 0 & -1 & 4 & -1 \\ 0 & 0 & 0 & 0 & 0 & -1 & 0 & 0 & -1 & 4 \end{matrix}$$

Fig. 1.3

The matrix E(10,4)

Test matrices of class $F2(m,n,c,r,\alpha)$ are $m \times n$ matrices which can be viewed as generalizations of the matrices of class D but with a lower left 10×10 triangle of elements added. $r-1$ is the width of a band located at a distance c from the main diagonal (and reappearing cyclicly under it). The elements are given by

$$a_{i,i-qn} = 1, \qquad i = 1\,(1)\,m;$$

$$a_{i,i-qn+c+s} = (-1)^s \cdot s \cdot i, \qquad s = 1\,(1)\,r-1, \quad i = 1\,(1)\,m;$$

where $q = 0,1,\ldots,\lceil m/n \rceil$ is chosen such that $1 \leq i - qn \leq n$ resp. $1 \leq i - qn+c+s \leq n$, and $\lceil m/n \rceil$ is the smallest integer greater than or equal to m/n;

$$a_{i,n-11+i+j} = j \cdot \alpha, \qquad i = 1\,(1)\,11-j, \quad j = 1\,(1)\,10;$$

$$a_{n-11+i+j,j} = i/\alpha, \qquad j = 1\,(1)\,11-i, \quad i = 1\,(1)\,10;$$

where $m \geq n \geq 22$, $11 \leq c \leq n-11$, $2 \leq r \leq \min(c-9,\ n-20)$, and $\alpha \geq 1$.

The smallest matrices of this class are thus $F2(22, 22, 11, 2, \alpha)$.

In Fig. 1.4 and 1.5 we show the sparsity pattern of matrices F2(26, 26, 12, 3, α) and F2(80, 30, 12, 4, α).

```
x o o o o o o o o o o o x x o o x x x x x x x x x x
o x o o o o o o o o o o o x x o o x x x x x x x x x
o o x o o o o o o o o o o o x x o o x x x x x x x x
o o o x o o o o o o o o o o o x x o o x x x x x x x
o o o o x o o o o o o o o o o o x x o o x x x x x x
o o o o o x o o o o o o o o o o o x x o o x x x x x
o o o o o o x o o o o o o o o o o o x x o o x x x x
o o o o o o o x o o o o o o o o o o o x x o o x x x
o o o o o o o o x o o o o o o o o o o o x x o o x x
o o o o o o o o o x o o o o o o o o o o o x x o o x
o o o o o o o o o o x o o o o o o o o o o o x x o o
o o o o o o o o o o o x o o o o o o o o o o o x x o
o o o o o o o o o o o o x o o o o o o o o o o o x x
x o o o o o o o o o o o o x o o o o o o o o o o o x
x x o o o o o o o o o o o o x o o o o o o o o o o o
o x x o o o o o o o o o o o o x o o o o o o o o o o
x o x x o o o o o o o o o o o o x o o o o o o o o o
x x o x x o o o o o o o o o o o o x o o o o o o o o
x x x o x x o o o o o o o o o o o o x o o o o o o o
x x x x o x x o o o o o o o o o o o o x o o o o o o
x x x x x o x x o o o o o o o o o o o o x o o o o o
x x x x x x o x x o o o o o o o o o o o o x o o o o
x x x x x x x o x x o o o o o o o o o o o o x o o o
x x x x x x x x o x x o o o o o o o o o o o o x o o
x x x x x x x x x o x x o o o o o o o o o o o o x o
x x x x x x x x x x o x x o o o o o o o o o o o o x
```

Fig. 1.4

Sparsity pattern of matrices F2(26, 26, 12, 3, α)

We emphasize here that by varying the parameters for test matrices of class F2 we can change the size n, the ratio m/n, the density NZ/n^2, the sparsity pattern, and the stability properties of the matrices and therefore carry out a rather systematic investigation of how the performance of a sparse matrix code depends on these quantities.

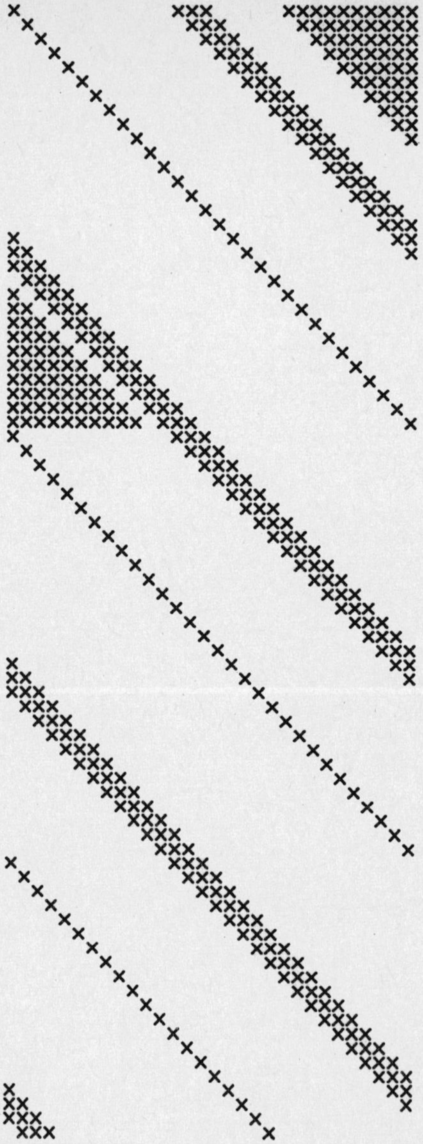

Fig. 1.5

Sparsity pattern of matrices F2(80, 30, 12, 4, α)

Table 1.6 below summarizes the dimension, the number of elements, and the smallest and largest elements of the test matrices.

The matrices of class D(n,c) and the matrices of class E(n,c) have been used by Zlatev in [92] and [96]. The matrices of class F2(m,n,c,r,α) have been used in [95] and [109]. Some details about the subroutines which generate matrices of these three classes are given in [104].

In addition to matrices of these three classes we have also used some Harwell test matrices (see [24]) in our numerical experiments. Some matrices arising in the discretization of certain chemical problems have also been used.

class	dimension	NZ	$\min\|a_{ij}\|$	$\max\|a_{ij}\|$
D(n,c)	n	4n + 55	1	max(1000,n+1)
E(n,c)	n	5n - 2c - 2	1	4
F2(n,n,c,r,α)	n	r · n + 110	1/α	max(rn-n,10α)

Table 1.6

Various characteristics of the test matrices

1.4 An example

To demonstrate the assertions at the end of section 1.2 we have solved a linear system with the coefficient matrix E(1000,44) (see section 1.3) using DS with the subroutines FO1BRE and FO4AXE from [64] and using IR with the package Y12M ([103], [108]).

For this matrix we have

$$n = 1000, \quad n^2 = 1000000 \quad NZ = 4910.$$

Details of the computations are summarized in the following table

Algorithm	storage	time	accuracy
	COUNT	in secs	$\lVert x - \tilde{x} \rVert$
DS	45850	152.31	2.02 E-1
IR	14082	8.50	1.83 E-6

Table 1.7

Storage, time and accuracy for the solution of a linear system with coefficient matrix E(1000,44). IR is used with T = 0.01 and 16 iterations (see the end of this section).

In this example (and in the following ones) we have chosen the right-hand-side such that the solution x is the vector consisting of 1's.

Note that this problem is very large if we are to solve it by conventional dense matrix techniques and even if the band structure is exploited we will need about 88000 storage locations for the solution process. Using the sparse matrix techniques which we are going to discuss in chapters 2 and 3 the space requirements can be cut by half (which also happens to be the saving if symmetry had been exploited), but the real gain is obtained with the techniques from chapter 4: iterative refinement + a large drop tolerance.

When new elements (fill-ins) are generated in the elimination process they are checked against the drop tolerance, T, and if they are smaller than T they are simply ignored. In this way we save space and computing time, but we also introduce large errors. In order to regain the accuracy we perform iterative refinement and, as seen from table 1.6, we actually get a better solution with IR than with DS.

1.5 Contents of chapters 2 - 5

In chapter 2 we shall describe a storage technique based on ordered lists and following the ideas of [51], [52], and we shall compare it with another technique using linked lists.

Chapter 3 is devoted to pivotal strategies focusing on the well-known Markowitz strategy ([57]) and some generalizations ([92]).

In chapter 4 we shall discuss drop tolerance and iterative refinement and show how to combine these into an algorithm which can be much more efficient than DS.

The techniques described in chapters 2 - 4 can also be used in more general problems where matrix A is rectangular, and with other solution methods. In chapter 5 we define a general computational scheme which includes many well-known and commonly used methods as special cases. Then we discuss briefly the use of sparse matrix techniques, pivoting, drop-tolerance and iterative refinement for the general scheme.

It should be mentioned here that the following chapters are based on the results obtained in [92], [93], [95], [96], [103].

Chapter 2: Storage Techniques

2.1 Input requirements

Assume that the matrix A is large and sparse. We shall not make assumptions on any particular structure of the elements of A. If such information is available (if e.g. A is positive definite or has a band structure) then it may be possible to take advantage of it and arrive at a more efficient computational scheme, but we shall focus our attention on more general techniques.

If the structure of the matrix allows us to use some iterative method, then the subroutines given in [91] could be successfully used. Some band matrix solver should be used if the elements of the matrix form a narrow band around the main diagonal. Such solvers can be found e.g. in the NAG Library [64] and in LINPACK [16]. It should be mentioned here that a condition number estimation can optionally be calculated (using the algorithm given in [14]; see also [35], [60] and [13]) when the LINPACK subroutines are used. The condition number estimation for band matrices is discussed also in [50]. The symmetry of the matrix is exploited in two well-known packages: SPARSPAK (see [43],[44]) and the Yale package (see [28], [29], [30], [31]). It should be mentioned that there are subroutines for general matrices in the Yale package ([29]). Exploiting the sparsity of symmetric matrices by the use of multifrontal techniques is described in [26]. For other techniques, which can be used with sparse matrices of special structure, see e.g. [20].

When no special structure is present every element of the coefficient matrix must be accompanied by information on where it belongs, i.e. in addition to the value of a_{ij} we must know the row number, i, and the column number, j. This information can be arranged in three one-dimensional arrays A, CNR, RNR containing the values a_{ij}, j, and i respectively. (If integers take as much space in our computer as reals do, then we must already at this point have $NZ < n^2/3$ in order to save space; we shall see later that even stricter bounds should be imposed on NZ.)

In general we cannot expect that the order in which the user wishes to supply the matrix-elements can be used effectively in the further computations, so in order to stay user-friendly we place no restrictions on this order. Any order will do, and we shall take care of restructuring the elements in a suitable way (see the next section).

Example 2.1

Consider the matrix (n = 5, NZ = 12)

$$(1.1) \qquad A = \begin{bmatrix} 5 & 0 & 0 & 3 & 0 \\ 2 & 4 & 0 & 0 & 1 \\ 0 & 1 & 3 & 0 & 2 \\ 0 & 0 & 0 & 2 & 3 \\ 0 & 0 & 0 & 2 & 1 \end{bmatrix}$$

In Fig. 2.1 we illustrate the use of the arrays A, CNR and RNR. Note that the length of array RNR (NN1) is less than the length of arrays A and CNR (NN). We shall see in the next section why this is so. Matrix A is rather small and not sparse according to our 'definition', but we use it here only as an illustration. ∎

	1	2	3	4	5	6	7	8	9	10	11	12 ... 20 .. 24
Real array A	5	4	3	2	1	3	1	2	3	2	1	2
Integer array CNR	1	2	3	4	5	4	5	5	5	1	2	4
Integer array RNR	1	2	3	4	5	1	2	3	4	2	3	5

Fig. 2.1

Contents of arrays A, CNR and RNR corresponding to matrix A

2.2 Reordering the structure

We shall now reorder the elements of A to get a structure which is practical to use with Gaussian elimination. This structure amounts to an ordering of A by rows and we shall describe two ways of accomplishing this.

We shall need four one-dimensional arrays (length n) of pointers. For practical reasons these are collected as columns in a two-dimensional array HA, and as we shall need seven more later on the array HA is declared to be $n \times 11$.

The pointers to be used here are

HA(i,1) : Number of elements with row number less than i.

HA(i,3) : Number of elements in row i (stage 1) / pointer to next element in row i (stage 2).

HA(j,4) : Number of elements with column number less than j.

HA(j,6) : Number of elements in column j (stage 1) / pointer to next element in column j (stage 3).

We shall return to the use of these pointers in section 2.3.

The first reordering process is done in three stages:

Stage 1. Make a copy of the elements of A and CNR in positions NZ + 1 to NZ + NZ of A and CNR. (Therefore we must have $NN \geqq 2 \cdot NZ$ with this process.) Count the number of elements in each row and place it in HA(·,3) and count the number of elements in each column and place in HA(·,6). Compute the total number of elements with row numbers less than i and place it in HA(i,1) and HA(i,3). Also compute the total number of elements with column numbers less than j and place it in HA(j,4) and HA(j,6).

Stage 2. Copy the elements of A (and CNR) in positions NZ + 1 to NZ + NZ back into the first NZ positions but ordered by rows using HA(i,3) as a pointer to where the next element in row i shall go.

Stage 3. In array RNR we store the row numbers of the matrix elements ordered by columns. More specifically, in positions HA(j,4) + 1 to HA(j+1,4) we store the row numbers of the elements of column j in matrix A.

```
      DO 20 I = 1, N
        PIVOT(I) = 0
   20   HA(I,3) = HA(I,6) = HA(I,1) = HA(I,4) = 0
C                        count number of elements in each row and column
      DO 30 I = 1, NZ
        J = CNR(I)
        CNR(NZ+I) = J
        A(NZ+I) = A(I)
        HA(J,6) = HA(J,6) + 1
        J = RNR(I)
   30   HA(J,3) = HA(J,3) + 1
      N1 = N - 1
C                        find the beginning of each row and column
      DO 40 I = 1, N1
        HA(I+1,1) = HA(I,1) + HA(I,3)
        HA(I+1,4) = HA(I,4) + HA(I,6)
        HA(I,3) = HA(I,1)
   40   HA(I,6) = HA(I,4)
      HA(N,3) = HA(N,1)
      HA(N,6) = HA(N,4)
C                        copy the elements back into  A
      DO 50 I3 = 1, NZ
        I = RNR(I3)
        I2 = HA(I,3) + 1
        I1 = NZ + I3
        CNR(I2) = CNR(I1)
        A(I2) = A(I1)
   50   HA(I,3) = I2
C                        store row numbers in  RNR
      DO 70 I = 1, N
        J1 = HA(I,1) + 1
        J2 = HA(I,3)
        DO 70 J3 = J1, J2
          J = CNR(J3)
          K = HA(J,6) + 1
          RNR(K) = I
   70     HA(J,6) = K
```

<u>Fig. 2.2</u>

FORTRAN code for the reordering

A	5	4	3	2	1	3	1	2	3	2	1	2	5	4	3	2	1	3	1	2	3	2	1	2
CNR	1	2	3	4	5	4	5	5	5	1	2	4	1	2	3	4	5	4	5	5	5	1	2	4
RNR	1	2	3	4	5	1	2	3	4	2	3	5												
HA(., 1)		0	2	5	8	10																		
HA(., 3)		0	2	5	8	10																		
HA(., 4)		0	2	4	5	8																		
HA(., 6)		0	2	4	5	8																		

A	5	3	4	1	2	3	2	1	2	3	1	2
CNR	1	4	2	5	1	3	5	2	4	5	5	4
RNR	1	2	2	3	3	1	4	5	2	3	4	5
HA(., 1)		0	2	5	8	10						
HA(., 3)		2	5	8	10	12						
HA(., 4)		0	2	4	5	8						
HA(., 6)		2	4	5	8	12						

<u>Fig. 2.3</u>

Contents of the arrays after stage 1 and stage 3

In Fig. 2.2 we give a FORTRAN implementation of this reordering and in Fig. 2.3 we give the contents of A, CNR, RNR and HA after stage 1 and stage 3.

We note that the contents of A, CNR, and HA(.,1) is enough to hold complete information on matrix A, i.e. $2 \cdot NZ + n$ locations are sufficient. In order to perform the elimination process more efficiently some extra storage (e.g. array RNR) is needed also after the input stage.

The code in Fig. 2.2 is just one way of restructuring the information, and it introduces the somewhat artificial condition that $NN \geqq 2 \cdot NZ$. Although the elimination process will often put harder conditions on NN, it might be instructive to look at another reordering process which needs no extra space in A and CNR.

This process can also be divided into three stages: stage 3 is identical to stage 3 in process 1, and so is stage 1 except for the copy of A and CNR.

In stage 2 we begin with picking out an element from A and reading its row number in RNR. Using HA(i,3) as a pointer to where the next element in row i should go we place our element there accompanied by its column number in CNR. But first we save the element which is already located there and the process can continue. The process will stop if we are to place an element where we picked out the first one. In order to postpone this event we start out with the element in position NZ (this is the location reserved for the last element encountered in row n). If the process stops before all elements have been placed we seek a new starting element among the positions reserved for the last element in row n - 1, n - 2, ... , using HA(.,1) as pointers. In order to discover that an element has been taken out from array A we need to set a flag. A negative number is placed in RNR for that purpose. As mentioned earlier we do not need the information in RNR after the sorting so we are not destroying useful information by placing -1's in RNR.

In Fig. 2.4 we give a FORTRAN implementation of this reordering which does not need extra space (except for the pointers in HA) and in Fig. 2.5 we give the contents of A, CNR and RNR after each of the three stages of the process. The code is slightly longer than for the first process but a closer examination reveals that it uses about the same number of operations. Anyway the bulk of the computation will most certainly lie somewhere else in the complete sparse matrix code.

Remark 2.2 If the matrix elements are already ordered by rows the first strategy will preserve the order whereas the second strategy will perform a cyclic permutation within each row. We can take advantage of the ordering by carrying out only stages 1 and 3 of the second process. ∎

It should be mentioned here that the reordering obtained by both strategies is based on ideas proposed in [51] and [52].

```
      DO 20 I = 1, N
   20   HA(I,3) = HA(I,6) = HA(I,1) = HA(I,4) = 0
C                           count number of elements in each row and column
      DO 30 I = 1, NZ
        J = CNR(I)
        HA(J,6) = HA(J,6) + 1
        J = RNR(I)
   30   HA(J,3) = HA(J,3) + 1
      N1 = N - 1
C                           find the beginning of each row and column
      DO 40 I = 1, N1
        HA(I+1,1) = HA(I,1) + HA(I,3)
        HA(I+1,4) = HA(I,4) + HA(I,6)
        HA(I,3) = HA(I,1)
   40   HA(I,6) = HA(I,4)
      HA(N,3) = HA(N,1)
      HA(N,6) = HA(N,4)
      I = RNR(NZ)
      J = CNR(NZ)
      XP = A(NZ)
      RNR(NZ) = -1
      K = N
C                           sort the elements of  A  and  CNR
      DO 50 I3 = 2, NZ
        I1 = HA(I,3) + 1
        HA(I,3) = I1
        I = RNR(I1)
        RNR(I1) = -1
        Z = A(I1)
        A(I1) = XP
        XP = Z
        J1 = CNR(I1)
        CNR(I1) = J
        J = J1
        IF(I.GT.0) GO TO 50
```

```
45     I2 = HA(K,1)
       K = K - 1
       I = RNR(I2)
       IF(I.LT.0) GO TO 45
       RNR(I2) = -1
       XP = A(I2)
       J = CNR(I2)
50   CONTINUE
     I1 = HA(I,3) + 1
     HA(I,3) = I1
     A(I1) = XP
     CNR(I1) = J
C                        store row numbers in  RNR
     DO 70 I = 1, N
       J1 = HA(I,1) + 1
       J2 = HA(I,3)
       DO 70 J3 = J1, J2
         J = CNR(J3)
         K = HA(J,6) + 1
         RNR(K) = I
70       HA(J,6) = K
```

<u>Fig. 2.4</u>

FORTRAN code for space-economic reordering

A	5	4	3	2	1	3	1	2	3	2	1	2
CNR	1	2	3	4	5	4	5	5	5	1	2	4
RNR	1	2	3	4	5	1	2	3	4	2	3	5
HA(.,1)		0	2	5	8	10						
HA(.,3)		0	2	5	8	10						
HA(.,4)		0	2	4	5	8						
HA(.,6)		0	2	4	5	8						

A	3	5	4	1	2	1	3	2	2	3	2	1
CNR	4	1	2	5	1	2	3	5	4	5	4	5
RNR	-1	-1	-1	-1	-1	-1	-1	-1	-1	-1	-1	-1
HA(.,1)		0	2	5	8	10						
HA(.,3)		2	5	8	10	12						
HA(.,4)		0	2	4	5	8						
HA(.,6)		0	2	4	5	8						

A	3	5	4	1	2	1	3	2	2	3	2	1
CNR	4	1	2	5	1	2	3	5	4	5	4	5
RNR	1	2	2	3	3	1	4	5	2	3	4	5
HA(.,1)		0	2	5	8	10						
HA(.,3)		2	5	8	10	12						
HA(.,4)		0	2	4	5	8						
HA(.,6)		2	4	5	8	12						

Fig. 2.5

Contents of the arrays after each of the three stages of the reordering

2.3 The elimination process

We are now ready to start the factorization or elimination process which, as mentioned in section 1.1, is performed in $n-1$ stages. Assume that we are about to begin the computations in stage k $(1 \leq k \leq n-1)$. The elements in row i of the coefficient matrix are located in positions HA(i,1) + 1 to HA(i,3) in array A with the column numbers given in CNR. It is also practical to know the locations of the elements of A_k (and of A_i for $i < k$). We therefore introduce the pointer HA(i,2) such that elements of A_k (or of A_i if $i < k$) are to be found in positions HA(i,2) + 1 to HA(i,3) of array A. We shall use the notation

$$(3.1) \qquad \begin{aligned} K_i &= HA(i,1) & \qquad \overline{K_j} &= HA(j,4) \\ L_i &= HA(i,2) & \overline{L_j} &= HA(j,5) \\ M_i &= HA(i,3) & \overline{M_j} &= HA(j,6) \end{aligned}$$

We have $K_i \leqq L_i \leqq M_i$ and $K_i = L_i$ at the beginning (see Fig. 2.6). Note that the elements in row i of the coefficient matrix are not ordered after column number to begin with, but we shall keep a partial ordering in the sense the elements in positions $K_i + 1$ to L_i have column numbers less than $\min(i,k)$ and those in positions $L_i + 1$ to M_i have column numbers larger than or equal to $\min(i,k)$. The column numbers of these elements are found in the same positions in array CNR. (See Fig. 2.6).

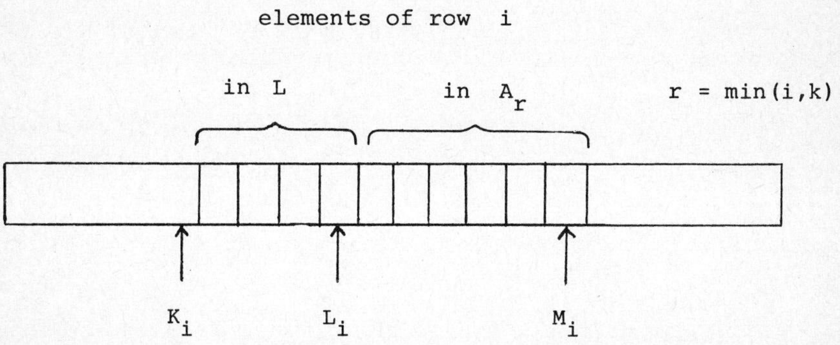

Fig. 2.6

The pointers K_i, L_i and M_i

Similarly in the column-ordered list, the row numbers less than k of elements in column j $(k \leqq j \leqq n)$ are found in position $\overline{K_j} + 1$ to $\overline{L_j}$ of RNR, and those greater than or equal to k are found in positions $\overline{L_j} + 1$ to $\overline{M_j}$ of RNR.

Remark 2.3 We keep array RNR in order to find the elements of a certain column easily. This is important when scanning A_k but the information is not needed for the first $k - 1$ columns of matrix A, and space in RNR can thus be freed for other use. Therefore the length of array RNR (NN1) can be smaller than NN. ∎

At the beginning of stage k in the elimination process the elements in row k with column numbers greater than or equal to k (i.e. locations $L_k + 1$ to M_k in array A) are copied into their proper places in the last $n - k + 1$ locations of a real array PIVOT (of length n) which has been initialized with 0's before stage 1. We assume that possible interchanges have been performed already (see chapter 3) such that $a_{kk}^{(k)}$, now located in PIVOT(k), is non-zero.

We shall now perform the calculations specified by formula (1.2.1) for those rows i for which $a_{ik}^{(k)} \neq 0$. These row numbers are found in RNR, in locations $\overline{L_k} + 1$ to $\overline{M_k}$.

For each such row, i, we first find the location of element $a_{ik}^{(k)}$ by searching through positions $L_i + 1$ through M_i of array CNR to find the value k. Interchange this element with the element sitting in location $L_i + 1$ (this affects A and CNR) and add 1 to L_i. Compute

$$t = a_{ik}^{(k)} / a_{kk}^{(k)} \tag{3.2}$$

and store in $A(L_i)$ (cf. section 2.6).

We now perform two sweeps:

a. Go through row i, locations $L_i + 1$ through M_i in CNR, and for each column number, j, check if $\text{PIVOT}(j) \neq 0$. If so change the corresponding element of A according to formula (1.2.1) which here reads

$$a_{ij}^{(k+1)} = a_{ij}^{(k)} - t \cdot a_{kj}^{(k)}, \tag{3.3}$$

and set $\text{PIVOT}(j) = 0$.

b. Go through row k to see if we have used all the elements, i.e. go through locations $L_k + 1$ to M_k in CNR and for each column number j check whether $\text{PIVOT}(j) = 0$.

If so, we just restore PIVOT(j) from A.

If not, a new element (fill-in) is created in row i according to formula (1.2.1) which now reads

$$a_{ij}^{(k+1)} = - t \cdot a_{kj}^{(k)}, \tag{3.4}$$

and we shall see in the next sections where to put it.

At the end of stage k we zero out the elements which we have used in PIVOT, locations $k + 1$ to n, such that PIVOT is ready for stage $k + 1$, but we keep $a_{kk}^{(k)}$ in PIVOT(k) as this can make the back-substitution faster.

Remark 2.4 The above description follows closely ideas given in [25], [68], [69]. ∎

2.4 Storage of fill-ins

New elements (fill-ins) are generated during the elimination whenever we use formula (3.4) and they should be stored in accordance with our general principles such that they can be treated during subsequent stages just like the 'old' elements of the coefficient matrix.

But first some good news. We have free space available in array A and CNR since we store the diagonal elements $a_{kk}^{(k)}$ elsewhere (in PIVOT(k)) and we have free space in array RNR since we do not need the information provided here for column numbers less than k. Free space is indicated by placing zeros in RNR and CNR.

We do not want free space in the middle of a row so unless $a_{kk}^{(k)}$ already occupies position M_k we interchange it with the element in posititon M_k, make this location free by setting CNR(M_k) = 0, and subtract 1 from M_k : M_k = HA(k,3) = $M_k - 1$.

A similar thing can be done in the column-ordered list (array RNR) but as mentioned already the whole of column k can be removed after stage k of the elimination is completed.

```
C                          is there room to the right
 165  IF(CNR(MI+1).GT.0) GO TO 170
C                          yes
        MI = MI + 1
        A(MI) = AIJ
        CNR(MI) = J
        HA(I,3) = MI
        IF(MI.GT.NREND)NREND = MI
C                          we are done
        GO TO 300
 170  KI = HA(I,1)
C                          is there room to the left
      IF(CNR(KI).GT.0) GO TO 180
C                          yes
        LI = LI - 1
        A(KI) = A(LI)
        A(LI) = AIJ
        CNR(KI) = CNR(LI)
        CNR(LI) = J
        HA(I,1) = KI - 1
        HA(I,2) = LI - 1
C                          we are done
        GO TO 300
 180  I2 = NREND - KI
C                          make a copy of row  I  at the end
 280    I3 = KI + 1
        DO 290 I3 = I3, MI
          A(I3+I2) = A(I3)
          CNR(I3+I2) = CNR(I3)
 290      CNR(I3) = 0
        LI = LI + I2
        HA(I,1) = NREND
        HA(I,2) = LI - 1
        HA(I,3) = NREND = MI = MI + I2 + 1
        A(NREND) = AIJ
        CNR(NREND) = J
 300    CONTINUE
```

Fig. 2.7

FORTRAN code for adding fill-ins to A and CNR

```
C                          record fill-in in the column-ordered list
      LJ = HA(J,5)
      MJ = HA(J,6)
C                          is there room at the bottom
 365  IF(RNR(MJ+1).GT.0) GO TO 370
C                          yes
        MJ = MJ + 1
        RNR(MJ) = I
        HA(J,6) = MJ
        IF(MJ.GT.N1END) N1END = MJ
C                          we are done
        GO TO 500
 370  KJ = HA(J,4)
C                          is there room at the top
      IF(RNR(KJ).GT.0) GO TO 380
C                          yes
        RNR(KJ) = RNR(LJ)
        RNR(LJ) = I
        HA(J,4) = KJ - 1
        HA(J,5) = LJ - 1
C                          we are done
        GO TO 500
 380  I2 = N1 END - KJ
C                          make a copy of column  J  at the bottom
 480  I3 = KJ + 1
      DO 490 I3 = I3, MJ
        RNR(I3+I2) = RNR(I3)
 490    RNR(I3) = 0
      HA(J,4) = N1END
      HA(J,5) = LJ + I2
      N1END = MJ + I2 + 1
      RNR(N1END) = I
      HA(J,6) = N1END
 500  CONTINUE
```

<u>Fig. 2.8</u>

FORTRAN code for adding fill-ins to RNR

We may thus have some free space available between rows (and columns) and whenever a fill-in is generated it might be a good idea to check the end or the beginning of the row (column) first.

If there is no such space we shall have to copy the whole row (column) into the free space in A and CNR (RNR) after the last used location.

This strategy is exemplified in the piece of FORTRAN code given in Fig. 2.7 for a fill-in of value AIJ in row I and column J.

NREND is the last used position in A and CNR, and LI and MI are supposed to hold the values

```
LI = HA(I,2) + 1
MI = HA(I,3).
```

The process for adding fill-ins to the column-ordered list is quite similar, but for completeness we provide the FORTRAN code in Fig. 2.8.

<u>Remark 2.5</u> Two strategies are now possible:

<u>a</u>. Whenever a fill-in is generated it is added to the row-ordered list and the column-ordered list before we continue ([103], [108]).

<u>b</u>. We perform two sweeps: First simulate the elimination column by column and add possible fill-ins to the column-ordered list. Next eliminate for real, row by row, computing new elements of A_{k+1} and storing fill-ins in the row-ordered list ([51]).

The advantage of strategy <u>b</u> is that all fill-ins in one column (row) are added to the column- (row-) ordered list in succession, so that we need to make at most one copy of the column (row) at any stage, and we are not liable to run out of space too soon. The disadvantage is that two sweeps are necessary. ∎

Example 2.6 Consider the matrix from example 2.1 and assume that the structure is ordered as described in section 2.2 (Fig. 2.5). Assume that no interchanges are made in the first stage of the elimination. A fill-in is produced as $a_{24}^{(2)} = -1.2 \neq 0$.

There is no free space at the end of the second row, but there is one empty location at the beginning because we have stored the diagonal element in PIVOT(1). So we move $l_{21}^{(1)}$ back one step, place $a_{24}^{(2)}$ in its place and set the pointers L_2 = (HA(2,2) =) $L_2 - 1$, K_2 = (HA(2,1) =) $K_2 - 1$. In the column-ordered list there is no free space around column four so a copy must be made at the end of the list.

The contents of the arrays after stage 1 is shown in Fig. 2.9. ∎

A	3	.4	-1.2	1	4	1	3	2	2	3	2	1				
CNR	4	1	4	5	2	2	3	5	4	5	4	5				
RNR	2	0	2	3	3	0	0	0	2	3	4	5	1	4	5	2
HA(.,1)		0	1	5	8	10										
HA(.,2)		0	2	5	8	10										
HA(.,3)		1	5	8	10	12										
HA(.,4)		0	2	4	12	8										
HA(.,5)		0	2	4	13	8										
HA(.,6)		1	4	5	16	12										

Fig. 2.9

Contents of the arrays after stage 1 of the elimination

Example 2.7 Consider the matrix and the structure after example 2.6. Assume that no interchanges are made at the second stage of the elimination. A fill-in is produced as $a_{34}^{(3)} = 0.3 \neq 0$. Now there is free space at the beginning of the third row in the row-ordered list and at the end of the fourth column in the column-ordered list. The contents of the arrays after the fill-in is shown in Fig. 2.10. ∎

A	3	.4	-1.2	1	.25	.3	3	1.75	2	3	2	1				
CNR	4	1	4	5	2	4	3	5	4	5	4	5				
RNR	0	0	3	0	3	0	0	0	2	3	4	5	1	2	5	4	3
HA(.,1)		0	1	4	8	10											
HA(.,2)		0	2	5	8	10											
HA(.,3)		1	4	8	10	12											
HA(.,4)		0	2	4	12	8											
HA(.,5)		0	2	4	14	9											
HA(.,6)		1	3	5	17	12											

Fig. 2.10

Contents of the arrays after stage 2 of the elimination

2.5 Garbage collections

There is a limit to how many copies we can make at the end of each of the lists, but if we are hitting against the upper limit of the arrays we have probably made several copies along the way and left free locations behind. (If not, then the matrix is not as sparse as we thought and the program should return a message asking for more space.) What is needed now is to compress the structure collecting all free locations into one connected set which can be used for future copies. In computer science this kind of process is often called 'garbage collection'.

Array RNR can be and should be treated separately from A and CNR, because the need for garbage collections probably will occur at different times. We shall describe the compression or garbage collection for RNR.

We cannot expect the columns to be ordered since we have copied intermediate columns to the end of the list several times. Instead of sorting the elements of say HA(·,4) we put a marker at the beginning of each column giving the number of the column. This is done by going through HA(j,4), j = k(1)n, placing here the row number of the first element of the column, and placing -j in RNR instead (see Fig. 2.11 which corresponds to example 2.7 with NN1 = 16 such that a garbage collection is necessary).

RNR	0	0	-2	0	-3	0	0	0	-5	3	4	5	-4	2	5	4
HA(·,4)		0	3	3	1	2										

Fig. 2.11

Contents of array RNR and HA(·,4) before garbage collection

We now go through RNR(i), i = 1(1)N1END (the last used position). If RNR(i) = 0 the place is free and we go on. If RNR(i) < 0, say -j, we are at the beginning of a column of elements (column number j) and we update the pointers HA(j,k), k = 4, 5, 6. IF RNR(i) ≠ 0 the element in position i should be copied to the first free location in the new list we are making. Fig. 2.12 gives the FORTRAN code for this compression algorithm and Fig. 2.13 gives the (similar) code for compression in the row-ordered list.

It is of course expensive to perform garbage collections too often. One way to avoid this is to work with large arrays, i.e. to choose large values of NN and NN1. But we must keep a certain balance between storage and computation time so the result will usually be a compromise and we must learn to live with some garbage collections. Furthermore we do not know the amount of fill-in beforehand except in very special situations so the values of NN and NN1 must be chosen largely by intuition or previous experience.

It should be mentioned in this connection that the program must check whether the garbage collection resulted in enough free space for the operations to continue and if not return a message to the user stating the problem and asking for more space.

The codes given in Fig. 2.12 and Fig. 2.13 are based on ideas used in the package MA28 [18].

```
C                        garbage collection in column-ordered list
C                        set up markers at the beginning of each column
      DO 410 I2 = K, N
        KJ = HA(I2,4) + 1
        HA(I2,4) = RNR(KJ)
 410    RNR(KJ) = - I2
      J2 = KJ = 1
      DO 450 I2 = K, N
C                        step through  RNR  until a new column starts
        DO 420 J2 = J2, N1END
 420      IF (RNR(J2).LT.0) GO TO 430
 430    IC = - RNR(J2)
        I3 = J2 - KJ
        RNR(J2) = HA(IC,4)
        MIC = HA(IC,6) - I3
C                        copy a column
        DO 440 J3 = KJ, MIC
          J4 = J3 + I3
 440      RNR(J3) = RNR(J4)
        HA(IC,4) = KJ - 1
        HA(IC,5) = HA(IC,5) - I3
        HA(IC,6) = MIC
        KJ = MIC + 1
        J2 = KJ + I3
 450  CONTINUE
      DO 460 J4 = KJ, N1END
 460    RNR(J4) = 0
      N1END = MIC
```

<u>Fig. 2.12</u>

FORTRAN code for garbage collection in the column-ordered list

```
C                        garbage collection in row-ordered list
C                        set up markers at the beginning of each row
      DO 210 I2 = 1,N
        KI = HA(I2,1) + 1
        HA(I2,1) = CNR(KI)
 210    CNR(KI) = - I2
      J2 = KI = 1
      DO 250 I2 = 1, N
C                        step through  A, CNR  until a new row starts
        DO 220 J2 = J2, NREND
 220      IF(CNR(J2).LT.0) GO TO 230
 230    IC = - CNR(J2)
        I3 = J2 - KI
        CNR(J2) = HA(IC,1)
        MIC = HA(IC,3) - I3
C                        copy a row
        DO 240 J3 = KI, MIC
          J4 = J3 + I3
          A(J3) = A(J4)
 240      CNR(J3) = CNR(J4)
        HA(IC,1) = KI - 1
        HA(IC,2) = HA(IC,2) - I3
        HA(IC,3) = MIC
        KI = MIC + 1
        J2 = KI + I3
 250  CONTINUE
      DO 260 J4 = KI, NREND
 260    CNR(J4) = 0
      NREND = MIC
      LI = HA(I,2) + 1
      MI = HA(I,3)
```

Fig. 2.13

FORTRAN code for garbage collection in the row-ordered list

2.6 On the storage of matrix L

When solving linear equations with a dense coefficient matrix it is an automatic procedure to store the elements of matrix L because space is available in the lower triangular part of A. When several sets of equations with the same coefficient matrix are to be solved, maybe one after another, computation time can be saved using the LU factorization, but in any case no extra time or space is needed for the storage of L.

With sparse matrices the situation is different. The matrices are often large and we shall generally reserve so little space for them that some garbage collections are performed during the factorization process. In this case we can save space, i.e. even less storage need to be reserved, or we can save time on garbage collections, if we do not retain L. Whenever an element below the diagonal is eliminated the space occupied by it is freed and can be used e.g. to store a fill-in. Even if a copy of the row still needs to be made we only copy the elements above the diagonal, and when a garbage collection is performed the structure can be compressed more tightly than before because only elements above the diagonal are considered. Also the computation time is reduced (slightly) because fewer elements have to be handled.

On the other hand, if several systems are to be solved, one after another, it is probably a good idea to retain L if at all possible, the extra space being compensated by a sizable reduction in the computation time. We shall return to this in the next section and in chapter 4.

In Table 2.14 we show the reduction of storage, measured by the value of COUNT (see p. 5), for some matrices of classes D(n,c) and E(n,c) with n = 1000. It is seen that a reduction in storage of 25 to 40% is obtained for these test matrices by not storing L.

c	Matrices of class D(n,c)			Matrices of class E(n,c)		
	with L	without L	%	with L	without L	%
4	8719	5564	64	8126	6128	75
44	16131	9823	61	27658	14289	52
84	16263	9724	60	21411	11123	52
124	16734	9902	59	17456	9934	57
164	16277	9803	60	14621	8602	59
204	15319	9625	63	12111	7575	63

Table 2.14

Comparison of the storage needed in the elimination of test-matrices depending on whether L is stored or not.

2.7 Classification of problems

A problem which requires the solution of one or more systems of linear algebraic equations belongs to one of the following 5 categories:

(1) $Ax = b$ — One system is to be solved.

(2) $Ax_r = b_r$ — Several systems with the same coefficient matrix are to be solved.

(3) $A_r x_r = b_r$ — Several systems of the same structure are to be solved, (see definition 2.8 below).

(4) $A_1 x_{r1} = b_{r1}$, $A_2 x_{r2} = b_{r2}$, $\vdots$ — Many systems of the same structure are to be solved. Furthermore the same coefficient matrix appears successively several times.

(5) $Ax = b$, $By = c$, $\vdots$ — Several systems with different coefficient matrices of different structure are to be solved.

<u>Definition 2.8</u> Two matrices A_1 and A_2 are said to have the same structure if their elements occupy the same positions, i.e. $a_{ij}^{(1)} \neq 0 \iff a_{ij}^{(2)} \neq 0$. ∎

<u>Remark 2.9</u> We shall also call the matrices $A_1, A_2, \dots, A_r, \dots$ of the same structure even if some of the elements become zero for certain values of r. ∎

The question of which sparse matrix technique is efficient depends to a large extent on the category of the problem which we shall see now.

Category (1) and (5): The lower triangular matrix, L, need not be stored and we can profit from the saving of space by declaring our arrays A and CNR smaller. Another alternative would be to keep the sizes of arrays A and CNR and expect not to waste very much time on garbage collections.

Category (2): The lower triangular matrix, L, is computed and stored when the first system is solved and all (the subsequent) systems are solved by substitution using the computed LU-factorization. Quite often the computation time for solving $x_1 = QU^{-1}L^{-1}Pb$ is only a small percentage of the computation time for the factorization (just like for dense matrices) and we can save considerably by keeping L.

Category (3): L need not be stored, but we can still use some of the information obtained during the first factorization such that during the subsequent eliminations we can

A. avoid searching for pivots (see chapter 3).
B. minimize the number of garbage collections.
C. cut down on the number of copies of rows/columns.

Category (4): Same as for category (3) except that L should be stored just as with category (2).

We shall see later that categories (2) and (4) are the most important ones from our point of view.

Returning to category (3) (and (4)), in order to avoid searching for pivots we keep information about the row and column interchanges performed during the first factorization in two n-dimensional arrays (columns 7 and 8 in HA can be used).

This requires no extra work since the information is needed anyway for the solution of the first system of equations: $x_1 = QU^{-1}L^{-1}Pb$, the row interchanges in order to perform the same interchanges in the right-hand-side (this could be done together with the elimination, however) and the column interchanges in order to sort out the unknowns in the right order before returning the solution.

Remark 2.10 A word of caution is needed here. Because of numerical instability we do not want to allow very small elements as pivots (and certainly not zeros) but as the elements of the matrices A_r are allowed to vary in size this might happen for one value of r even if it didn't for $r = 1$. Therefore we must keep an eye on the pivots and possibly readjust the pivotal sequence once in a while. The introduction of a drop tolerance (see section 4.2) confuses the picture even more. ∎

Let r_i be the maximum number of elements in row i at any stage of the elimination process and let c_j be the maximum number of elements in column j likewise. Define

$$R = \sum_{i=1}^{n} r_i \quad ; \quad C = \sum_{j=1}^{n} c_j .$$

These values can be computed after the first system of equations has been solved.

If we reserve space for our arrays A, CNR and RNR such that $NN \geq R$ and $NN1 \geq C$ then the storage in both the row-ordered list and the column-ordered list can be arranged such that at the subsequent eliminations no copies of rows or columns need be made and no garbage collections are necessary.

If either $NN < R$ or $NN1 < C$ or both then some copies of rows or columns or both must be made and we can probably not avoid garbage collections either. The optimum size of the arrays involves a compromise between storage space and computation time and must be determined in practice for each particular problem and depending on the computer installation.

2.8 A comparison of ordered and linked lists

So far we have discussed one storage technique based on ordered lists. Another technique which was very popular in the sixties is based on the so-called linked lists. We shall use the matrix from example 2.1 to show the basic ideas behind this technique. Again three large arrays are necessary (one real and two integer arrays; we shall use the names A, CNR and RNR as before) and there is no reason to give them different lengths (i.e. NN = NN1). Two extra integer arrays of length n are needed pointing to the first element in each row and column (we shall use HA(·,1) and HA(·,4)).

As illustrated in Fig. 2.15 the contents of array RNR is the location of the next element in the same row. Corresponding to the last element in a row one places a number larger than NN in RNR and it is customary to use NN + the row number. In order to find the row number of a given element in array A (if we don't know it beforehand) we have to search through the list until we reach the last element in the row and then subtract NN from the contents of RNR. This is clearly a cumbersome way unless the matrix is very sparse and stays that way.

	1	2	3	4	5	6	7	8	9	10	11	12	13	14	15	16	17	18	19	20
real array A	5	4	3	2	1	3	1	2	3	2	1	2								
integer array CNR	10	11	23	6	7	12	8	9	25	21	22	24	14	15	16	17	18	19	20	-1
integer array RNR	6	7	8	9	12	21	10	11	24	22	23	25	14	15	16	17	18	19	20	-1

	1	2	3	4	5
HA(.,1)	1	2	3	4	5

	1	2	3	4	5
HA(.,4)	1	2	3	4	5

Fig. 2.15

The array CNR is used in a completely similar way with respect to the columns, see Fig. 2.15 for details.

Remark 2.11 Although we have used the words 'first', 'next' and 'last' we do not assume the elements or the linkage between them to be ordered within the rows/columns. The 'first' element in a row is just the element which happens to be the first one in our linked list. ∎

A code based on these ideas is MA18 [15], but we shall now outline an extension which can be useful if we do not store the matrix L or we use a large drop tolerance or we store the diagonal elements elsewhere (in array PIVOT). In these cases we shall generate free locations in arrays A, CNR and RNR and we might as well put them to use. We therefore link all the unused locations of arrays A, CNR and RNR together to form the so-called "free list" which can be used for storage of fill-ins. If locations are freed during the elimination process they can be added to the free list. The only extra thing needed is a pointer to the beginning of the free list (in Fig. 2.15 the free list begins in location 13).

And now for a comparison of the two storage techniques.

A. Reordering of the structure.

This is easier to do with linked lists, since no reordering of the elements in A is necessary. The computation time will be less than half of that for the ordered lists, but this part of the program takes a very small part of the time anyway.

B. Arithmetic operations and search for pivots.

Many operations involve finding the column (row) number of an element in a given row (column). As already noted this is a tedious process with linked lists unless there are very few non-zero elements in the matrix at all stages of the elimination. This is the main drawback with linked lists and maybe the only one, but it is a serious one.

C. Storage of fill-ins.

This is easy to do with linked lists. To add a new element in row i and column j amounts to taking the first element from the free list and tie up the links accordingly. No copies and no garbage collections are ever needed.

D. Storage space.

When working with linked lists it is not necessary to reserve more space in the arrays than what is actually needed for the elimination process and in this respect the situation resembles the one which we described in the last paragraph of section 2.7 for problems of category (3) and (4). But in general the ordered lists need some extra 'elbow room' for making copies such that we don't spend all our time making garbage collections. An example showing how the garbage collections and the total computing time can depend on the 'elbow room' is given in table 2.16. It must be mentioned, however, that array RNR must have length NN when using linked lists but can be considerably shorter with ordered lists and thus part of the savings is used again. It should also be mentioned that we usually do not know beforehand how much space is needed, and it is therefore difficult to take full advantage of this nice property with the linked lists.

Nowadays it is believed that the draw-back of B overshadows the advantages of A, C and D, a belief which is strengthened by practical work during recent years. But the world is neither completely white nor completely black and the choice between the two storage techniques depends on the programming language and the compiler as well as the problem. E.g. if we know that the matrix is very sparse and stays that way then we should prefer linked lists to ordered lists.

A program based on linked lists is MA18 [15]. Programs based on ordered lists are MA28 [18] and Y12M [103], [108].

NN = COUNT + s·n	T = 0.0, COUNT = 3474			T = 0.1, COUNT = 1994		
s	number of garbage coll.	computing time	per-cent	number of garbage coll.	computing time	per-cent
> 15	0	1.12	100	0	.48	100
6	11	1.37	122	3	.54	113
5	12	1.33	119	5	.56	117
4	16	1.42	127	7	.54	113
3	19	1.45	129	9	.62	129
2	25	1.55	138	16	.65	135
1	43	1.77	158	29	.76	158

Table 2.16

Dependence of garbage collections and computing time on elbow room for two runs with a test matrix of class F2 with $n = 100$, $NZ = 1110$ and $NN = COUNT + s \cdot n$. The significance of the drop tolerance T is mentioned in chapter 4.

Chapter 3: Pivotal Strategies

3.1 Why interchange rows and columns?

When doing Gaussian elimination it is necessary to make sure that $a_{kk}^{(k)} \neq 0$, since we should like to divide by that number. When dealing with dense matrices it is customary to interchange rows and/or columns such that not only is $a_{kk}^{(k)} \neq 0$ but it is the largest element in absolute value in column k of A_k, or in row k of A_k, or in the whole of A_k.

When dealing with sparse matrices we should like to relax this requirement because we also have another objective when performing row and column interchanges: minimization of fill-in. We shall therefore select a real $u \geqq 1$ and only require that

$$(1.1) \qquad u \cdot \left|a_{kk}^{(k)}\right| \geqq \left|a_{ik}^{(k)}\right|, \qquad i = k+1(1)n, \quad \text{or}$$

$$(1.2) \qquad u \cdot \left|a_{kk}^{(k)}\right| \geqq \left|a_{kj}^{(k)}\right|, \qquad j = k+1(1)n, \quad \text{or}$$

$$(1.3) \qquad u \cdot \left|a_{kk}^{(k)}\right| \geqq \left|a_{ij}^{(k)}\right|, \qquad i,j = k(1)n$$

corresponding to partial pivoting with row interchanges, partial pivoting with column interchanges, or complete pivoting, respectively.

It is desirable to keep u small for reasons of numerical stability. If b_k denotes the maximum element in absolute value of $A^{(k)}$ then we have for partial pivoting

$$(1.4) \qquad b_n \leqq (u+1)^{n-1} \cdot b_1 .$$

The quantity b_n enters into the a priori estimates [67] of the magnitude of the elements of the perturbation matrix E in (1.1.10) which we would like to keep rather small.

We should not be too afraid of using a somewhat large value of u, however, and for several reasons. Although the bound (1.4) can be attained for matrices of a special structure ([86]) it is not a realistic estimate for practically occurring matrices. (If it were, then even $u = 1$ would mean disaster for large n). For sparse matrices the number of non-zero elements in a column should replace n in the exponent of (1.4) ([37]) - and even this is not realistic. And at last we can note that the actual values of b_k can be computed and checked against a 'safety-factor' as the elimination takes place such that we can be warned if the growth of the elements is too large.

For complete pivoting a much lower bound than (1.4), but still rather pessimistic and unrealistic, can be obtained ([83], [86]). But the work involved in checking all of A_k at each stage is great and is generally not compensated by better stability or accuracy of the results.

A reasonably robust and reliable code can be based on partial pivoting provided we check the growth of elements in $A^{(k)}$, and check for small pivot elements in order to detect near-singularity of A.

<u>Remark 3.1</u> There are examples of matrices that are nearly singular without ever producing small pivot elements ([86]) and such pathological cases will remain undetected. ∎

In what follows we shall assume that $u > 1$ such that we still have a choice in selecting the pivot element and we shall utilize this choice to minimize the fill-in. We shall not attempt to find a strategy that will lead to the smallest possible amount of fill-in for the whole elimination. This would necessitate a very extensive and expensive search procedure and is completely unrealistic. We shall not even take much pains to find the element which produces the least fill-in in the computational stage which we are about to begin. Firstly, this pivotal strategy would not necessarily lead to the smallest over-all fill-in and, secondly, the search would still be rather expensive. What we shall do is generalize and improve on a pivotal strategy which was first suggested in [57], a strategy which is easy to implement, and which usually produces an amount of fill-in which, although probably not minimal, is small enough for the over-all procedure to be efficient.

3.2 The Markowitz strategy

Assume that the first $k-1$ stages of the Gaussian elimination have already been performed and that we are about to find the k'th pivotal element. Let $r(i,k)$ denote the number of non-zero elements in row i of A_k, and let $c(j,k)$ denote the number of non-zero elements in column j of A_k. A_k is defined in chapter 1 as the lower right $(n-k+1) \times (n-k+1)$ submatrix of $A^{(k)}$ and is called the 'active part' of matrix $A^{(k)}$. Its rows (columns) are the active parts of the rows (columns) of $A^{(k)}$.

Definition 3.2 The Markowitz cost of element $a_{ij}^{(k)}$ is

(2.1) $$M_{ijk} = (r(i,k)-1)\cdot(c(j,k)-1), \qquad (i,j = k(1)n). \qquad \blacksquare$$

M_{ijk} is equal to the number of matrix-elements which will change value from $A^{(k)}$ to $A^{(k+1)}$ if $a_{ij}^{(k)}$ is chosen as pivotal element, and is thus an upper bound for the amount of fill-in which can be produced if we choose $a_{ij}^{(k)}$. Let

(2.2) $$M_k = \min\{M_{ijk} \mid i,j = k(1)n\}.$$

The original Markowitz strategy amounts to, at any stage k, choosing a pivotal element with Markowitz cost M_k. This will not necessarily mean that we minimize the amount of fill-in at stage k, but it is considerably easier to compute the Markowitz cost than to compute the amount of fill-in for each element in A_k, and in practice it is almost as good (cf. numerical experiments in [69]).

There are (at least) two drawbacks with the Markowitz strategy:
1. There are still many elements in A_k to search through; and
2. We may encounter instability.

In order to limit the search Curtis and Reid ([15]) have in MA18 ordered the rows and the columns after increasing number of non-zero elements and the search may often be stopped rather quickly (see section 3.5).

Objection no. 2 points to the fact that very small elements can be selected as pivots with destructive effects on the numerical significance of the results. The answer to this is that our pivoting strategy must be a compromise somewhere between maximum stability and minimum fill-in.

3.3 The generalized Markowitz strategy (GMS)

In order to preserve numerical stability we shall not accept very small elements as pivots but instead introduce a stability factor $u \geqq 1$ as mentioned in section 3.1 and insist that formula (1.1) (or possibly (1.2) or (1.3)) be fulfilled.

In order to reduce the amount of search we shall not look at the whole submatrix A_k, but only consider a certain number, p, of rows from it, selected such that we have a good chance of keeping the amount of fill-in down close to the minimum.

Remark 3.3 When $p > 2$ and $k > n-p+1$ then A_k contains less than p rows so in order to be more precise we can state that we shall look at $\min(p,n-k+1)$ rows at stage k. ∎

We define a set of row numbers

(3.1) $$I_k = \{i_s \mid s = 1(1)\min(p,n-k+1), \qquad k \leqq i_s \leqq n\},$$

with increasing values of $r(i,k)$, i.e.

(3.2) $$i_s \in I_k \wedge i_t \in I_k \wedge s < t \Rightarrow r(i_s,k) \leqq r(i_t,k),$$

and containing the smallest values of $r(i,k)$:

(3.3) $$i_s \in I_k \wedge i \notin I_k \wedge k \leqq i \leqq n \Rightarrow r(i_s,k) \leqq r(i,k).$$

Furthermore we define the sets:

(3.4) $$B_k = \{a_{ij}^{(k)} \in A_k \mid |a_{ij}^{(k)}| \cdot u \geqq \max_{k \leqq m \leqq n} (|a_{im}^{(k)}|), \quad i \in I_k\},$$

(3.5) $$C_k = \{a_{ij}^{(k)} \in B_k \mid M_{ijk} = M_k'\},$$

(3.6) $$M_k' = \min\{M_{ijk} \mid a_{ij}^{(k)} \in B_k\}.$$

The elements of C_k are the candidates for pivotal elements. They satisfy a stability condition and at the same time have minimum Markowitz cost among a certain subset of elements from A_k. We should point out here that we may have rejected elements as candidates, not on account of violation of our stability requirements, but because they happen to be located in rows which we don't look at. Although this can happen, particularly when p is small, it is not likely to happen too often because we have selected the rows with the smallest number of non-zero elements and we would expect to find elements with low Markowitz cost here.

Definition 3.4 The generalized Markowitz strategy (GMS) amounts to choosing any element of C_k as pivotal element at stage k of the Gaussian elimination. ∎

Remark 3.5 The original Markowitz strategy corresponds to GMS with $u = \infty$ and $p = n$. ∎

Various values of u and p have been used or recommended in published programs as shown in Table 3.1 where rec. means that the particular value is recommended and not obligatory.

Author(s)	year	code	u	p
Curtis & Reid [15]	1971	MA18	4 rec.	n
Duff [18]	1977	MA28	10 rec.	n
Zlatev, Barker, Thomsen [98]	1978	SSLEST	[4,16] rec.	3 rec.
Zlatev & Thomsen [105]	1976	ST	[4,16] rec.	2

Table 3.1

Used or recommended values of u and p in various codes

It should be mentioned that the parameters u and p in Y12M are the same as in SSLEST (see [108]).

In order to investigate the effect of using very small values of p versus a large one we have performed an experiment using 20 Harwell test-matrices ([24]) and the values $p = 1$, 3 and n. The code SSLEST ([98]) has been used with $p = 1$ and 3 and the code MA28 ([18]) has been used with $p = n$. The use of two different codes will of course introduce obscuring side-effects and thus make the experiment less than optimal but we would not have done complete justice to the case $p = n$ by just selecting this value of p in SSLEST because sorting of the rows is not necessary in this case. We have chosen MA28 because this is a program designed for the case $p = n$ and considered a very efficient program, and would thus provide a fair basis for deciding on the best value of p.

In Table 3.2 we give the total memory requirement (measured by the sum of the values of COUNT - cf. p. 5) for all 20 systems and we note that the total COUNT is only increased by 7.3% when going from $p = n$ to $p = 1$, but the total computing time is reduced by about 50%. The intermediate value of $p = 3$ looks like a fine compromise with about the same computing time but only half the increase in COUNT.

p	total COUNT	%	total time
1	71322	107.3	31.33
3	68836	103.5	33.35
n	66491	100.0	61.92

Table 3.2

Dependency of COUNT and time on p for 20 Harwell test matrices

Remark 3.6 It must be emphasized here that some of the individual matrices showed much larger variation in COUNT - up to about 26% either way (see [92] for more details), so we should be careful not to pretend that we can draw general conclusions from Table 3.2. We must point out that some classes of sparse matrices may be rather sensitive to a reduction in p and take this information into account when deciding on a strategy (code) for our particular problem. (It is no great help for me that a code is 3% better on the average if it is 25% worse on my problem.) ∎

3.4 The improved generalized Markowitz strategy (IGMS)

Early experiments with GMS showed certain problems with numerical instability and we shall reproduce two of them here in order to see the problem and how it can be remedied.

Example 3.7 Consider the matrix, [92],

$$(4.1) \qquad A = \begin{bmatrix} 1+a & -v & o & o & o & o \\ o & 1+a & -v & & o & o \\ o & o & 1+a & -v & o & o \\ \vdots & \vdots & & \ddots & \ddots & \vdots \\ o & o & o & \cdots & 1+a & -v \\ v & o & o & \cdots & o & 1+a \end{bmatrix}$$

where $v > 1 + a$ and $a > 0$ is chosen close to the machine accuracy ε. Since all rows and columns contain only two non-zero elements the pivotal strategy is independent of the value of p for this matrix. If $v \leqq u$ then the elements on the main diagonal may be chosen as pivots and in this case

$$a_{nn}^{(n)} = 1 + a + v^2/(1+a) + \ldots + v^n/(1+a)^{n-1} .$$

This means that the GMS may cause unstable results (and even overflows) when n is large. Note, however, that no instability takes place if we always choose the largest element among the candidates for pivots. ∎

Example 3.8 Consider the matrix $A = E(2500, 50)$. Using MA28 on an IBM 3033 at NEUCC (Northern Europe University Computing Centre in Lyngby, Denmark) we have calculated $b_n = 2.2\,E7$. The corresponding solution was quite wrong. Note that if we always choose the largest element as pivotal, then we shall only pick elements on the main diagonal (see Remark 3.14) and we shall find $b_n = 4$ indicating perfectly stable computations. The system with matrix $A = E(2500, 50)$ has also been solved by the use of the subroutines of package Y12M (see [103], [108]). Some characteristics obtained during the solution process are given in Table 3.3 in order to illustrate the fact that the statements made at the end of the previous example are also valid for matrices of class E(n,c). It must be noted too that for this class of matrices the IGMS (see Definition 3.10 below) implemented in Y12M preserves the sparsity much better than the GMS which is applied in MA28. Of course, this leads to a great reduction of the computing time. It should be mentioned that all components of the solution vector are equal to 1. ∎

	MA28	Y12M (L kept)	Y12M (L removed)
Computing time	502.29	29.39	28.97
Storage (COUNT)	186453	88709	54145
Value of b_n	2.2 E7	4.0	4.0
Accuracy	2.1 E1	8.8 E-4	8.2 E-4

Table 3.3

Remark 3.9 Whereas the matrix in Example 3.7 is artificial, the matrices of class E(n,c) are very similar to matrices that appear in the numerical solution of certain elliptic differential equations and Example 3.8 therefore raises a serious objection against the GMS and calls for an improvement; see [92]. ∎

Definition 3.10 The improved generalized Markowitz strategy (IGMS) amounts to choosing as pivot the largest in absolute value among the candidates in C_k at stage k of the Gaussian elimination. ∎

Since all candidates for pivots have the same Markowitz cost we might as well use stability considerations when selecting one of them as pivotal element and this is the proposed 'improvement'. Note that any pivotal sequence which can result from applying IGMS can also be obtained by using the GMS. We cannot guarantee that IGMS in general is better than GMS, i.e. produces more stable computations, although there is good computational evidence for it. For special classes of matrices we can, however, prove the superiority of IGMS.

Theorem 3.11 If matrix A is diagonally dominant and symmetric in structure, then Gaussian elimination is stable when any IGMS is used.

Remark 3.12 It is well-known (see e.g. [84]) that pivoting for stability is not necessary for these matrices, but we might still want to make interchanges in order to preserve sparsity. ∎

Proof Let $1 \leqq k \leqq n-1$ and assume that only diagonal elements have been chosen as pivots in the first $k-1$ stages of the Gaussian elimination. By this choice the symmetry in structure ($a_{ij} \neq 0 \Longleftrightarrow a_{ji} \neq 0$) is preserved, and the active part of matrix A at stage k, A_k, is diagonally dominant too ($|a_{ii}^{(k)}| > \sum_{j \neq i} |a_{ij}^{(k)}|$). Therefore

$$c(j,k) = r(j,k), \qquad j = k(1)n,$$

and

$$M_{ijk} = (r(i,k)-1) \cdot (r(j,k)-1).$$

Let

$$r(i_1,k) = r(i_2,k) = \ldots = r(i_s,k), \qquad (1 \leqq s \leqq p).$$

Then the diagonal elements in rows $i_1, i_2, \ldots, i_s$ are elements of C_k and the largest element of C_k is one of these elements and will thus be chosen as pivot at stage k by any IGMS independently of the stability factor u.

Since this holds for any k ($1 \leqq k \leqq n-1$) an induction argument shows that only diagonal elements will be chosen as pivots in the elimination, and Wilkinson's analysis [84], [85], [86] gives $b_n \leqq 2 \cdot b_1$ indicating stability. ∎

Example 3.13 That pivoting to preserve sparsity can be necessary is shown by the matrix with sparsity pattern given below

$$\begin{matrix} x & x & x & x & \ldots & x \\ x & x & o & o & \ldots & o \\ x & o & x & o & \ldots & o \\ x & o & o & x & \ldots & o \\ \vdots & & & & \ddots & \vdots \\ x & o & o & o & \ldots & x \end{matrix}$$

If no pivoting is performed then we shall have complete fill-in, i.e. the sparsity is completely destroyed. If on the other hand the matrix is diagonally dominant and any IGMS is employed no fill-ins appear. ∎

Remark 3.14 The diagonal dominance requirement can be slightly relaxed [90, p. 37]. A closer review of the proof of Theorem 3.11 and of Wilkinson's analysis ([84]) reveals that we need only

$$|a_{ii}| \geq \sum_{j \neq i} |a_{ij}| \qquad i = 1(1)n$$

together with strict inequality for at least one value of i, and

$$|a_{ii}| > \max_{j \neq i} |a_{ij}| \qquad i = 1(1)n.$$

The test matrices of class $E(n,c)$ are diagonally dominant in this weaker sense. ∎

Theorem 3.15 If matrix A is symmetric and positive definite, then Gaussian elimination is stable when any IGMS is used provided $u = \infty$ if p is large enough.

Proof The proof follows the same lines as the proof of Theorem 3.11. Since $u = \infty$ all elements in the $s \times s$ submatrix formed by taking rows and columns $i_1, i_2, \ldots, i_s$ out of $A^{(k)}$ are also elements of C_k, and as this submatrix is positive definite its largest element lies on the diagonal and will be chosen as pivot at stage k by the IGMS. ∎

Example 3.16 The matrix

$$A = \begin{Bmatrix} 1 & 0 & 900 \\ 0 & 10 & 50 \\ 900 & 50 & 900\,000 \end{Bmatrix}$$

is positive definite. If $u < 5$ then $a_{13} = 900$ will be chosen as pivot at stage 1 in the Gaussian elimination and the interchanges will destroy the symmetric structure. If $u > 5$ and $p \geq 2$ then we shall choose $a_{22} = 10$ (or if $u > 900$ and $p = 1$

we shall choose $a_{11} = 1$) as pivot, thus preserving stability. We may choose relatively small elements as pivots but as shown by Wilkinson [84] this will not violate the stability of the computations with a positive definite matrix. ∎

All the pivotal sequences which can result from using IGMS are also possible pivotal sequences for GMS (with the same p and u). But if the GMS just once selects a pivotal element outside the diagonal, then the symmetric structure is lost and we cannot guarantee stability. As Example 3.8 implies, this does happen in practice. Note also that the problems we encountered in Example 3.7 with the GMS are eliminated with the IGMS.

It is possible to identify qualities with a matrix which indicate that IGMS is a better strategy than GMS. If there are candidates for pivots at several stages of the elimination which are about as small as allowed by the stability factor, u, then the probability of selecting one of those is great and we shall often find $b_{k+1} \approx (1+u) \cdot b_k$ with the GMS; see [92].

Example 3.8 shows that this can have detrimental effects and in order to make a more thorough investigation we have made calculations with 96 matrices of class E(n,c) with n = 250(50)1000 and c = 4(40)204. The routine we have selected to represent the GMS is the NAG subroutine F01BRE (see also [18] and [25]) with the recommended value of u = 10 for the elimination, and the NAG subroutine F04AXE for the substitution. The representative of the IGMS was Y12M ([103], [108]) also with u = 10, but the value of u is immaterial for these matrices when IGMS is used. The computations were performed on the UNIVAC 1100/82 computer at RECKU (the Regional Computing Centre at Copenhagen University) using single precision ($\varepsilon \simeq 1.5$ E-8). In no cases were the routines using GMS better, but for about 25% of the matrices the decomposition by F01BRE was very inaccurate and so were the results calculated by F04AXE. The right-hand sides were chosen such that the vector consisting of 1's was always the solution.

In Table 3.4 we give the computation time, the value of COUNT and the accuracy for a typical value of n (n = 800). It is seen that Y12M is better in all cases and in every respect, and in particular the gain is great for 'intermediate' values of c (here 44 and 84). Note that a typical value in practice would be $c = \sqrt{n} \simeq 28$. In Table 3.5

we have chosen c = 44 and compared the accuracy for various n. In the last column iterative refinement (see chapter 4) has been used.

c	F01BRE + F04AXE			Y12M		
	Time	COUNT	Accuracy	Time	COUNT	Accuracy
4	2.71	9420	5.49 E-4	2.45	6504	1.00 E-8
44	53.67	30424	4.13 E-1	6.88	9882	5.96 E-8
84	24.73	22868	1.31 E-2	4.43	8793	1.12 E-6
124	12.45	16778	2.36 E-3	3.67	7849	7.45 E-8
164	7.69	13951	1.29 E-3	3.15	7218	1.49 E-8
204	6.06	12166	7.05 E-5	2.48	6443	2.98 E-8

Table 3.4

Comparison of NAG routines with Y12M for matrices of class E(800,c). Y12M is used with T = 0.01 and the IR option in this test

n	F01BRE + F04AXE	Y12M - DS	Y12M - IR T = 0.01
650	1.42 E-2	1.06 E-5	2.98 E-8
700	2.87 E-2	1.24 E-5	1.49 E-8
750	1.03 E-2	1.38 E-4	2.98 E-8
800	4.13 E-1	1.60 E-5	5.96 E-8
850	4.74 E-3	1.79 E-5	2.25 E-7
900	7.34 E-1	2.50 E-5	9.24 E-7
950	2.61 E-2	2.65 E-5	1.83 E-6
1000	2.02 E-1	2.98 E-5	8.24 E-7

Table 3.5

Comparison of the accuracy obtained by the NAG routines with the accuracy obtained by Y12M (DS and IR option) for matrices of class E(n,44).

3.5 Implementation of the pivotal strategy

In this section we shall discuss some of the practical problems connected with the implementation of the pivotal strategy. We shall describe in detail the strategy used in the subroutine Y12M ([103], [108]) which is based on IGMS with a small value of p and compare it with MA28 ([18]) which uses GMS with $p = n$.

We shall search the p (or rather $\min(p, n-k+1)$) rows with smallest numbers of non-zero elements in stage k of the Gaussian elimination.

Rather than going through all $n-k+1$ rows of A_k to find the p 'best' we shall order the rows after increasing number of elements and use the p 'first'. Three arrays of length n are needed for the efficient storage and handling of this information, one array to hold the information and two arrays to update it. In Y12M three columns of the integer array HA (see section 2.2) have been used: columns no. 7, 8, and 11. Column HA(·,7) holds the row numbers ordered after increasing number of elements. This information must be updated after each stage of the elimination because we must remove the pivotal row, remove the elements in the pivotal column, and add fill-ins. In HA(i,8) we store the position of row i in the ordered list HA(·,7), and in HA(j,11) we keep the position (in HA(·,7)) of the first row with j elements. If there is no such row we set HA(j,11) = 0.

By using these three arrays ($3 \cdot n$ locations) we can in a very efficient manner keep track of the number of non-zero elements in the rows of A_k and keep the rows ordered accordingly since only a few rows are altered during one stage of the elimination. Therefore the number of operations is $O(n \cdot c)$ where c is the average number of rows per stage rather than $O(n^2)$ if a search was to be performed every time.

Furthermore the first k positions of HA(·,7) and HA(·,8) (and the last k positions of HA(·,11)) are not needed after stage k of the elimination and they can therefore be used to store information about the row and column interchanges.

All elements in the p 'best' rows are investigated in order to find the elements of the set C_k. Therefore a small value of p ($p \leq 3$) is recommended. The largest element of C_k is selected, i.e. IGMS is implemented in Y12M, but our pivotal strategy is easy to modify e.g.

such that diagonal elements could be chosen as pivots.

The use of columns 7, 8, and 11 of HA is illustrated in Fig. 3.6 where we show the contents at the beginning of the Gaussian elimination on the matrix from Example 2.1, and in Fig. 3.7 which gives the situation after stage 1, supposing that element (1,1) was chosen as pivot.

	1	2	3	4	5
HA(·,7)	4	5	1	3	2
HA(·,8)	3	5	4	1	2
HA(·,11)	0	1	4	0	0

Fig. 3.6

Contents of columns 7, 8, and 11 of HA before stage 1 for the matrix from Example 2.1.

	1	2	3	4	5
HA(·,7)	1	5	4	3	2
HA(·,8)	1	5	4	3	2
HA(·,11)	0	2	4	0	0

Fig. 3.7

Contents of columns 7, 8, and 11 of HA after stage 1 for the matrix from Examples 2.1 and 2.6.

For reasons of comparison we shall briefly look at the routine MA28 to see how the implementation can be done with a large value of p.

Again it is not efficient to search the whole of A_k in order to find out what the minimum Markowitz cost is (if we can avoid it). It is better to search the rows in order of increasing number of non-zero elements, but this time also the columns must be searched. Therefore extra integer arrays are needed to keep the columns ordered, too. The rows and columns are now searched in order of increasing number of elements, and if there are rows and columns with the same number of elements, the rows are investigated first. Each element is checked against the stability criterion (1.2) - therefore it is much easier to check elements rowwise than columnwise - and the Markowitz cost M_{ijk} is computed.

If the row (or column) currently being searched has s non-zero elements in its active part and the best M_{ijk} so far is smaller than or equal to $(s-1)^2$ (or $s(s-1)$) then the search can be terminated because no element in the remaining part of A_k can have a smaller Markowitz cost.

In this way the amount of search can be somewhat reduced. Duff [18] p. 25 reports that an average of 14 rows and columns were searched in a matrix with $n = 199$.

The implementation in MA28 corresponds to a GMS and it is probably very inefficient to adapt it to an IGMS. The element which is selected as pivot in MA28 is just one element from C_k - in order to find the largest one we must keep searching the rows (or columns) with s elements until $(s-1)^2$ (or $s(s-1)$) becomes greater than M_k', and not just $\geq$. This could increase the amount of search considerably, particularly if the matrix contains many rows/columns with the same (small) number of elements.

Our comparison of the two strategies thus leads to the result that our strategy (from Y12M) uses less time for the pivotal search (because fewer rows are searched, and only rows are searched), uses less space (because no arrays are needed to keep track of the columns), and possibly leads to more stable computations (because an IGMS is implemented). On the other hand we can expect more fill-ins since we search very few (2 - 3) rows and therefore may not select a pivotal element with minimum Markowitz cost.

Practical experiments indicate, however, that the increase in COUNT is usually very slight and does not disturb the overall efficiency of our scheme.

3.6 Other strategies

Much of the work (i.e. computation time) and extra space connected with the various implementations of variants of the Markowitz strategy is spent searching through, reordering, and keeping track of rows and columns. Therefore it is tempting to suggest yet another pivotal strategy which minimizes this work. This strategy proceeds as follows:

Order the columns of matrix A after increasing number of non-zero elements. At stage k of the Gaussian elimination choose as pivot an element in column k which satisfies a stability condition and in which row there is a minimal number of non-zero elements. Since we expect to find rather few elements in column k we might choose to check them all and not bother to keep the rows sorted. This pivotal search is much simpler than those we have considered so far and furthermore we perform only row interchanges during the elimination.

Numerical results in [21] (see also [19] p. 120) show some drawbacks with methods based on this strategy: They often produce many fill-ins, possibly because columns with few elements to begin with quickly are contaminated with fill-ins but are still used as pivotal columns. Therefore this strategy must be used with care and/or with special classes of matrices.

An interesting conjecture is that the sparsity might be preserved better if this pivotal strategy is combined with the use of a large drop-tolerance (and iterative refinement, see next chapter).

For certain classes of matrices it is possible to preserve both stability and sparsity by choosing pivotal elements only on the main diagonal - and in some cases they can even be used in the natural order, thus avoiding interchanges at all. Much work and extra bookkeeping can be avoided this way and it is therefore a good idea to furnish a sparse matrix code with options such that these special - but frequently occurring - cases can be dealt with efficiently.

Diagonally dominant matrices and positive definite matrices are two classes for which no pivoting is needed in order to preserve stability. Still, pivoting might be advantageous in some cases to preserve sparsity. In this connection it should be mentioned that the use of a large drop-tolerance (and iterative refinement) often can reduce the amount of fill-in, such that these simpler pivotal strategies can be used to advantage; see [103], [108].

Some other strategies are discussed in [21]; see also [19] and [73].

Chapter 4: Iterative Refinement

4.1 Convergence of iterative refinement

Recall from chapter 1 that the coefficient matrix of the linear system

(1.1) $$Ax = b$$

is decomposed into

(1.2) $$LU = PAQ + E,$$

and that an approximation - the direct solution (DS) - x_1 to x is already calculated by

(1.3) $$x_1 = Q\,U^{-1}\,L^{-1}\,P\,b.$$

The iterative refinement (IR) is the process (see [36], [77], [85], [86])

$$\left.\begin{aligned} &(1.4) && r_i = b - Ax_i, \\ &(1.5) && d_i = Q\,U^{-1}\,L^{-1}\,P\,r_i, \\ &(1.6) && x_{1+1} = x_i + d_i, \end{aligned}\right\} \quad i = 1, 2, \ldots, q-1$$

which is terminated for some q for which

(1.7) $$\|x_q - x_{q-1}\| \leqq \varepsilon \cdot \|x_q\|,$$

or

(1.8) $$\|d_q\| > \|d_{q-1}\| \quad \wedge\ q > 2$$

or

(1.9) $$q = MAXIT.$$

ε denotes the machine accuracy, $\|\cdot\|$ is any vector norm, and MAXIT is a prescribed maximum number of iterations.

We have the following theorems about the convergence of the iterative refinement process.

Theorem 4.1   Let x be the true solution of (1.1) and assume that all computations with (1.4) - (1.6) are performed without errors. If

$$F = U^{-1} L^{-1} E \tag{1.10}$$

then

$$x_{i+1} - x = QF^iQ^T(x_1 - x) = -QF^{i+1}Q^T x, \qquad i = 0, 1, \ldots \tag{1.11}$$

and

$$d_{i+1} = QF^iQ^T d_1, \qquad i = 0, 1, \ldots \tag{1.12}$$

Proof   The proof is by induction and is left to the reader as an exercise. (Note that $F = I - U^{-1} L^{-1} PAQ$). ∎

Theorem 4.2   Let λ_k $(k = 1(1)n)$ denote the eigenvalues of F numbered such that

$$|\lambda_1| \geq |\lambda_k| \qquad (k = 2(1)n). \tag{1.13}$$

Under the same assumptions as in Theorem 4.1 we have

$$x = x_j + \sum_{i=j}^{\infty} d_i \qquad (j = 1, 2, \ldots) \tag{1.14}$$

if

$$|\lambda_1| < 1. \tag{1.15}$$

In the affirmative case

$$\lim_{i\to\infty} x_i = x \tag{1.16}$$

and

$$\lim_{i\to\infty} d_i = 0. \tag{1.17}$$

<u>Proof</u> The proof follows easily from Theorem 4.1. ■

<u>Corollary 4.3</u> The relations (1.14), (1.16), and (1.17) hold if (1.15) is replaced by

(1.18) $\|F\| < 1$,

where $\|\cdot\|$ is any matrix norm induced by the vector norm chosen. ■

<u>Remark 4.4</u> By (1.13) $|\lambda_1|$ is the spectral radius of matrix F. We shall use the notation.

(1.19) $\rho(F) = |\lambda_1|$. ■

<u>Remark 4.5</u> The number

(1.20) $\text{RELEST} = \|d_{q-1}\|/\|x_q\|$

is called the estimated relative error. ■

We can now take a closer look at the three stopping criteria (1.7) - (1.9) in the light of the theorems.

If $\rho(F) \ll 1$ then the iterative process (1.4) - (1.6) will converge quickly and will typically be stopped by relation (1.7) or possibly (1.8) if rounding errors get to dominate. RELEST will give a good estimate of the relative error in the computed solution, x_q (if $b \neq 0$).

If $\rho(F) < 1$ but close to one the rate of convergence may be very slow and the iteration will typically be stopped by (1.9). RELEST will provide a fair error estimate.

If $\rho(F) \geqq 1$ then the IR will probably not converge and the relation (1.17) will not hold. This is detected by (1.8), normally with $q = 3$, and the value of RELEST will tell what happened.

We must emphasize that the theorems hold under the assumption that the IR is performed without errors (an error analysis in which the errors made in the calculation of (1.3) - (1.6) are taken into account is given in [59] and [111]; see also [36]). This is of course not true in practice. The computation of the residual vectors, r_i, by (1.4) is normally carried out in extended precision and there is good experimental evidence that this is sufficient for the iterative process to converge to within machine accuracy (given (1.15)). In this connection it can be mentioned that the computational errors made in the substitution (1.5) are usually much smaller than those made in the decomposition (see [77], [83], [85] and [86].

4.2 The drop tolerance

When comparing iterative refinement (IR) with direct solution (DS) it is immediately seen that IR requires more space (because a copy of A must be held) and more computation time (for the process (1.4) - (1.6)). What we buy for this price is higher accuracy of the solution (if IR converges) and an error estimate. This might not be enough in many cases since the DS often gives sufficient accuracy and since error estimates can be obtained by other means (see e.g. [14], [35] and [16], see also section 4.8).

When dealing with sparse matrices the problem gets a new dimension, however. It is important for the efficiency to limit the amount of fill-in during the elimination. Now many of the fill-ins generated are rather small in magnitude compared with the original matrix elements. It is a natural thing to throw away fill-ins which are smaller than ε (the machine accuracy) times the original elements, but since errors are 'permitted' during the elimination (the matrix E) it is tempting also to disregard other small elements which appear at this stage ([110]).

To be more precise: we introduce a quantity, T, called the drop tolerance, and whenever a new element is generated by (2.3.4) and is less than T in absolute value we set it to 0. $T = 0$ corresponds to the ordinary situation, $T = \varepsilon \cdot b_1$ is natural choice, and a larger value, say $T = 0.01$ is what we propose.

To use a large value of T will cause the generation of rather large elements in the error matrix E in (1.2) and will reduce the accuracy of the direct solution (1.3), probably to the degree of unacceptability.

It is therefore necessary to regain the accuracy and for this the iterative refinement process is ideal. The computations involved in (1.4) - (1.6) are usually not very time-consuming compared to the triangular decomposition (1.2) such that 10 or 15 iterations can easily be performed. By Corollary 4.3 convergence is assured if matrix F is not too large. If we assume that the coefficient matrix A is scaled to have a norm close to 1 then by (1.10)

$$(2.1) \qquad \|F\| \leq \|U^{-1}L^{-1}\| \cdot \|E\| \simeq \kappa(A) \cdot \|E\|$$

where $\kappa(A) = \|A\| \cdot \|A^{-1}\|$ is the condition number of A.

The size of the elements of E depends on the magnitude of T, and relation (2.1) indicates that a large value of T can be used for well-conditioned matrices, but that T must be reduced to ensure convergence for ill-conditioned coefficient matrices.

We have already - in Example 1.4, chapter 1 - seen the effects of using IR with a large drop tolerance. A large reduction in storage, is in turn accompanied by a sizeable reduction in computing time, because fewer elements need to be treated in the elimination phase. The iterative refinement itself takes only 10% of the time (in this example). On top of this the accuracy is much better with IR, even though the DS was computed with T = 0.

Also in Table 2.16, chapter 2 we have seen good effects from using a large drop-tolerance: not only is the value of COUNT smaller, but with the same amount of elbow-room a large T seems to imply fewer garbage collections.

4.3 Storage comparisons

In this section we shall take a closer look at the storage requirements of a code using dense matrix technique (DMT) and two options of a code using sparse matrix technique (SMT-DS, SMT-IR). As a typical DMT code we are thinking of DECOMP/SOLVE ([35]) but there is only little variation between efficient DMT codes. Also in the SMT case our results are fairly general - we use the code Y12M which is the running example for these notes. We shall take NN1 = NN as the size of the arrays.

Now the space needed for the three cases is

(3.1) $S_1 = n^2 + 3n$, for DMT;

(3.2) $S_2 = 3 \cdot NN + 13n$, for SMT-DS;

(3.3) $S_3 = 2 \cdot NZ + 3 \cdot NN + 17n$, for SMT-IR.

Assume that

(3.4) $NN = \nu \cdot NZ$.

As suggested in the previous section we shall expect NN - and ν - to be smaller for SMT-IR than for SMT-DS, so when comparing these options we shall introduce indices on NN and ν.

From (3.1), (3.2), and (3.4) we have

(3.5) $S_2 < S_1 \iff \frac{NZ}{n^2} < g(\nu,n) = \frac{1}{3\nu}(1 - \frac{10}{n})$.

We have

(3.6) $g(\nu,n) < g(2,n) < \lim_{n\to\infty} g(2,n) = \frac{1}{6} \qquad (\nu > 2)$.

From (3.6) we read the following criterion

Criterion 4.6 If more than 1/6 of the elements of matrix A are non-zero then DMT will use less space than SMT-DS. ∎

It is difficult to formulate a converse of Criterion 4.6: a criterion for when SMT-DS is more space-efficient than DMT, because we usually do not know how much space will be needed for fill-ins during the elimination. If this information (COUNT) is available, then NN = COUNT + 2n is probably a good choice (see Table 2.16, chapter 2). The function $g(\nu,n)$ also depends on n, but for the large values of n which we expect, this variation is only small. Some selected values of $g(\nu,n)$ are shown in Table 4.1.

n \ ν	2	3	4	5	6	7
50	.133	.089	.067	.053	.044	.038
100	.150	.100	.075	.060	.050	.043
1000	.165	.110	.082	.066	.055	.047
∞	.167	.111	.083	.067	.056	.048

Table 4.1

The function $g(\nu,n)$

To compare DMT and SMT-IR we use (3.1), (3.3) and (3.4) to get

(3.7) $$S_3 < S_1 \Leftrightarrow \frac{NZ}{n^2} < g^*(\nu,n) = \frac{1}{2+3\nu} \cdot (1 - \frac{14}{n}).$$

The function g^* satisfies

(3.8) $$g^*(\nu,n) < g^*(2,n) < \lim_{n\to\infty} g^*(2,n) = \frac{1}{8},$$

which immediately gives

Criterion 4.7 If more than 1/8 of the elements of matrix A are non-zero then DMT will use less space than SMT-IR. ∎

Again it is difficult to express the converse: when does it pay to use a sparse matrix technique, but at least we can supply a table of typical values of the function $g^*(\nu,n)$ which behaves very similarly to $g(\nu,n)$.

n \ ν	2	3	4	5
50	.090	.065	.051	.042
100	.108	.078	.061	.051
1000	.123	.090	.070	.058
∞	.125	.091	.071	.059

Table 4.2

The function $g^*(\nu,n)$

Finally we should like to compare SMT-DS and SMT-IR w.r.t. space requirements. Instead of (3.4) we therefore assume

(3.9) $NN_2 = \nu_2 \cdot NZ, \quad NN_3 = \nu_3 \cdot NZ$

and expect that $\nu_3 < \nu_2$.

From (3.2), (3.3) and (3.9) we have

(3.10) $S_3 < S_2 \Leftrightarrow 2 + \frac{4n}{NZ} < 3(\nu_2 - \nu_3)$.

Assume that $NZ \geq 4n$ - probably a safe assumption. Then we have from (3.10)

Criterion 4.8 SMT-IR uses less space than SMT-DS provided $\nu_3 < \nu_2 - 1$, i.e. provided the IR option uses NZ fewer locations in each of the three major arrays than the DS option. ∎

A few words are needed concerning the assumptions ((3.4) and) (3.9) and the expected values of ν_2 and ν_3. Nothing general can be said but we have collected experimental results with our test matrices, the Harwell matrices ([24]), various practical problems of chemical origin ([70], [71]) and thermodynamical problems ([106], [107]). Typical values for ν_2 are in the range [3,5] if the matrix L is retained (see section 2.6) - if L is not needed then $\nu_2 \in [2,3]$ in many cases - but occasionally values up to 10 can be seen (Table 4.3). When IR is used we shall always assume a large value of the drop tolerance $(T \simeq 0.01)$ and $\nu_3 > 3$ occurs hardly ever. In Table 4.3 we compare an SMT-DS code (F01BRE + F04AXE from the NAG-library) with Y12M in IR-mode (T = 0.01) on the test matrices E(n,44) which typically generate many fill-ins. On this sort of matrices SMT-IR is especially effective.

If we can use $\nu_2 < 5$ or $\nu_3 < 3$ - a more realistic assumption for ν_3 than for ν_2 - then we can from Tables 4.1 and 4.2 deduce

Criterion 4.9 SMT will use less space than DMT provided less than 6% of the elements of matrix A are non-zero. ∎

Remark 4.10 Criterion 4.9. is rather conservative. In many cases we can allow 10% non-zeros and the sparse-matrix codes will still come out better. ∎

n	NZ	SMT-DS (F01BRE)		SMT-IR (Y12M)	
		COUNT	COUNT/NZ	COUNT	COUNT/NZ
650	3160	22246	7.04	7697	2.44
700	3410	24286	7.12	8453	2.48
750	3660	26932	7.36	9174	2.51
800	3910	30424	7.78	9882	2.53
850	4160	35290	8.48	11643	2.80
900	4410	37230	8.44	12360	2.80
950	4660	40488	8.69	12551	2.69
1000	4910	45850	9.34	14082	2.87

Table 4.3

Space comparison of SMT-DS and SMT-IR
(T = 0.01) on matrices of class E(n,44)

Remark 4.11 In formulae (3.2) and (3.3) we have assumed that NN1 = NN. As mentioned in section 2.1 we can often choose NN1 $\simeq$ 0.6 · NN and this will make sparse matrix codes even more efficient.

Also if integers take up less space than reals this will be to the advantage of sparse matrix codes. ∎

4.4 Computing time

When comparing the computing time of various linear equation solvers several factors play an important role. For dense matrix codes the situation is fairly straightforward: the computing time is directly proportional to the number of arithmetic operations, $\frac{1}{3} n^3$ - if an arithmetic operation is defined appropriately as one multiplication, one addition and three references to array elements, and if n is reasonably big.

For sparse matrix codes the following items are important:

The dimension of the matrix, n;
The number of non-zero elements, NZ;
The sparsity pattern;
The amount of fill-in; and
The pivoting strategy.

Even when the amount of fill-in is roughly the same the pivoting strategy itself can cause great differences in the computing time. Therefore we shall compare the NAG-routines (F01BRE and F04AXE) with the DS- and the IR-option of Y12M. F01BRE uses GMS and Y12M uses IGMS and the combined effects of IR (T = 0.01) and a better pivoting strategy is seen in Table 4.4 which contains the computing times corresponding to Table 4.3.

n	F01BRE - DS	Y12M - IR
650	23.47	4.55
700	28.46	5.27
750	38.29	6.12
800	53.67	6.88
850	58.54	6.97
900	57.35	7.47
950	115.58	8.07
1000	152.31	8.50

Table 4.4

Computing time in seconds on a UNIVAC 1100/82 for matrices of class E(n,44). IR is used with T = 0.01.

The matrices E(n,c) are symmetric and positive definite band matrices. We have tested Y12M against a routine specifically designed for such matrices but which does not take advantage of the sparsity: the NAG routine F04ACE ([88]).

n	c	F04ACE	Y12M - IR
900	30	5.17	8.15
961	31	5.90	8.84
1024	32	6.59	9.41
1089	33	7.40	10.02
1156	34	8.29	10.57
1225	35	9.25	11.23
1296	36	10.28	11.86
1369	37	11.41	12.65
1444	38	12.62	13.33
1521	39	13.93	14.23
1600	40	15.35	14.76
2304	48	30.66	21.45

Table 4.5

Computing time in seconds on a UNIVAC 1100/82 for matrices of class E(n,c). IR is used with T = 0.01.

It should be mentioned that F04ACE is most efficient on band matrices with narrow bands. Nevertheless it performs rather well on the matrices of class $E(n,\sqrt{n})$, but for larger values of n the sparse matrix code Y12M - which is written for general matrices and exploits neither symmetry nor positive definiteness - is superior.

We have performed a series of experiments in order to show the dependence of the computing time on n and the sparsity pattern and in order to compare the NAG routines with both the DS- and the IR-option of Y12M. In Table 4.6 we show the results for test matrices of class D(n,c), c = 4(40)204, n = 650(50)1000. The numbers are the sums of the computing times in seconds for the six values of c measured on the UNIVAC 1100/82 at RECKU.

n	F01BRE	Y12M - DS	Y12M - IR
		$T = 10^{-12}$	$T = 10^{-2}$
650	47.11	29.59	13.16
700	59.63	33.81	13.00
750	57.57	34.99	13.77
800	65.69	36.22	14.55
850	69.04	26.63	15.16
900	77.36	41.41	16.00
950	90.91	40.71	16.28
1000	82.92	42.13	17.85

Table 4.6

Computing times for matrices of class D(n,c), c = 4(40)204 in seconds on a UNIVAC 1100/82.

In Table 4.7 we give the similar numbers for matrices of class E(n,c).

n	F01BRE	Y12M - DS	Y12M - IR
		$T = 10^{-12}$	$T = 10^{-2}$
650	52.45	25.53	16.43
700	69.31	29.09	19.10
750	86.68	31.82	21.55
800	109.38	36.52	23.06
850	131.07	42.68	25.33
900	143.69	46.31	27.45
950	227.23	52.04	29.51
1000	253.87	58.62	31.79

Table 4.7

Computing times for matrices of class E(n,c), c = 4(40)204 in seconds on a UNIVAC 1100/82.

In Table 4.8 we show the dependence of the computing time on c for n = 800. Small and large values of c seem to be the 'easiest' to solve for all three codes but the difference is not great for Y12M-IR. Intermediate values of c are especially tough for F01BRE although they also put a certain strain on Y12M-DS and this is the reason for the great differences in the performance of the three codes as seen in Tables 4.6 and 4.7. But in all cases Y12M-DS is better than F01BRE and in turn Y12M-IR is better still, but for the exception that proves the rule (E(800,4)).

c	D(800,c)			E(800,c)		
	F01BRE	Y12M-DS $T = 10^{-12}$	Y12M-IR $T = 10^{-2}$	F01BRE	Y12M-DS $T = 10^{-12}$	Y12M-IR $T = 10^{-2}$
4	3.45	2.42	2.37	3.24	2.11	2.45
44	11.85	6.50	2.09	52.52	14.76	6.88
84	14.15	6.97	2.43	24.99	8.19	4.43
124	14.19	7.34	2.44	12.69	4.52	3.67
164	12.92	6.49	2.75	7.79	4.11	3.15
204	9.13	6.50	2.47	6.15	2.83	2.48

Table 4.8

Computing times for matrices D(800,c) and E(800,c) in seconds on a UNIVAC 1100/82.

In Table 4.9 we compare F01BRE with Y12M-IR on matrices of class F2(500, 500, 20, r, 100). By varying r (r = 5(5)40) we change the sparsity of the matrices, thereby making the problem harder. In all cases Y12M-IR is 3 - 5 times faster than the NAG-routine.

r	NZ	$\frac{NZ}{n^2}$	F01BRE	Y12M - IR
				$T = 10^{-2}$
5	2610	0.01	9.91	2.22
10	5110	0.02	32.96	6.16
15	7610	0.03	56.84	11.60
20	10110	0.04	59.32	14.84
25	12610	0.05	131.39	25.59
30	15110	0.06	97.69	34.32
35	17610	0.07	144.16	50.76
40	20110	0.08	288.03	62.81

Table 4.9

Computing times for matrices of class F2(500, 500, 20, r, 100) in seconds on a UNIVAC 1100/82.

4.5 Choice of drop tolerance and stability factor

We have seen that the use of iterative refinement together with a large value of the drop tolerance, T, gives very efficient computations. We have performed several experiments in order to find out just how large a T to choose. In Tables 4.10 and 4.11 we compare the DS-option $(T = 10^{-12})$ of Y12M with the IR-option and three different values of T.

It is seen from Tables 4.10 and 4.11 that a large value of the drop tolerance leads to a shorter computation time, and we have seen earlier that the space requirements are smaller and that the iterative refinement also gives better accuracy than the direct solution. The exact size of T is not very critical, but should of course be small enough for IR to converge.

A rule-of-thumb to use with matrices which are not too ill-conditioned and whose non-zero elements are of the same order of magnitude, a, is to choose $T \in [10^{-5} \cdot a, 10^{-2} \cdot a]$. With ill-conditioned matrices a smaller value of T may be needed in order to ensure convergence, cf. (2.1).

n	DS	IR		
	$T = 10^{-12}$	$T = 10^{-4}$	$T = 10^{-3}$	$T = 10^{-2}$
250	8.83	5.87	5.38	5.35
300	16.52	8.31	7.33	6.52
350	19.00	9.07	7.48	7.37
400	21.11	9.96	8.85	7.76
450	30.13	11.56	9.88	9.09
500	24.11	11.93	10.44	9.48
550	38.47	13.61	11.66	10.24
600	36.52	14.52	13.00	11.16
Total	194.69	84.83	74.02	66.97

Table 4.10

Sums of computing times for matrices of class D(n,c), c = 4(40)204, in seconds with the code Y12M on a UNIVAC 1100/82.

n	DS	IR		
	$T = 10^{-12}$	$T = 10^{-4}$	$T = 10^{-3}$	$T = 10^{-2}$
250	3.45	3.66	3.54	3.32
300	4.49	4.72	4.47	4.42
350	6.38	6.31	5.84	5.68
400	8.51	8.11	7.32	6.96
450	10.51	10.25	9.11	8.49
500	13.49	12.70	11.00	10.03
550	15.67	15.23	12.83	11.50
600	20.01	20.00	14.73	12.94
Total	82.51	80.98	68.84	63.34

Table 4.11

Sums of computing times for matrices of class E(n,c), c = 4(40)204, in seconds with the code Y12M on a UNIVAC 1100/82.

A special strategy can be used with problems of category 4 (see section 2.7) where many systems of the same structure are to be solved. In this case it might be profitable to set a large initial value of T (say T = a) and try to solve the systems. If a system cannot be solved to within a prescribed error tolerance (RELEST $< \varepsilon$) then decrease T by a factor, $T := c \cdot T$, $(c < 1)$, and solve again. With this strategy we accept some extra work in the beginning trying to find an optimal T and reduce the total work; see [72], [93], [96] and [107].

This strategy has been used on linear systems arising from the use of two-stage, modified diagonally implicit Runge-Kutta methods ([94]) on large systems of ordinary differential equations arising from chemical problems ([72]). We show the results of comparing this strategy with Y12M-DS and Y12M-IR in Table 4.12.

Algorithm	Strategy	initial T	final T	COUNT	iter	time
Y12M-DS	fixed T	10^{-14}	10^{-14}	24347	1	41.25
Y12M-IR	fixed T	10^{-2}	10^{-2}	17318	3.77	16.77
Y12M-IR	variable T	10^{0}	10^{-1}	13517	5.54	13.00

Table 4.12

Comparison of drop tolerance strategies on a chemical problem. iter is the average number of iterations and time is the average time for solving two systems. n = 255, NZ = 7715 and COUNT is the largest value encountered in any of the systems.

We have introduced a large T in order to limit the number of fill-ins. In chapter 3 we tried to achieve this by using a large stability factor, u, at the risk of instability. It is therefore interesting to investigate the combined effects of u and T. An experiment was carried out with a matrix of class F2 and the results are summarized in Table 4.13.

T	u = 4			u = 512		
	COUNT	iter	time	COUNT	iter	time
0	3376	7	5.68	3044	7	4.99
.01	1790	10	2.44	2218	11	3.47
.1	1475	12	2.25	1947	11	3.01
1	1120	13	1.79	1333	15	2.34
10	860	11	1.27	946	21	2.32

Table 4.13

The effect of u and T on the computational efficiency of Y12M-IR for the matrix A = F2(125, 125, 15, 6, 4) on a CDC Cyber 173.

It is seen from Table 4.13 that a large T gives smaller storage and computing time, independent of u, even though the number of iterations may increase. The effect of changing u is much smaller, but there is an indication that u should not be chosen too large when T is large. This somewhat surprising result may be due to the fact that rather small pivotal elements can be chosen when u is large, and small pivotal elements tend to produce fill-ins of large magnitude. So although we produce fewer fill-ins we retain most of them despite a large T. Table 4.13 only refers to a single matrix and no general results should be inferred from this alone, but several experiments with test matrices of class D(n,c) point in the same direction; see [96].

4.6 When and how to use iterative refinement

We have seen several examples where the use of iterative refinement - and a large drop tolerance - was superior to direct solution. Of course this is not true every time and three typical exceptions deserve mentioning.

(i) If matrix A is very ill-conditioned, more precisely: $\kappa(A) \cdot \varepsilon > 1$, then the iterative refinement may not converge. This condition is machine-dependent so one might switch to another machine (a CDC Cyber 173 has $\varepsilon \simeq 10^{-15}$ and this is sufficient in most cases), or compute a direct solution in double precision.

(ii) If matrix A is very large and storage requirements are very important, then we may not have room for L (see section 2.6) and an extra copy of A. On the other hand IR plus a large T usually implies much less fill-in and quite often IR pays anyway.

(iii) When the number of fill-ins is small in the first place there is not much to gain by using IR. We shall call such a problem a 'cheap' problem. A typical example of a very cheap problem is a dense-band matrix such as $E(n,2)$ which produces no fill-ins.

Using the model of section 4.3 we can estimate the extra amount of storage when using IR relative to DS as

$$\text{(6.1)} \qquad \frac{S_3 - S_2}{S_2} = \frac{2NZ + 4n}{3 \cdot \nu \cdot NZ + 13n}$$

Since $NZ \geqq n$ and $\nu \geqq 1$ we have

$$\text{(6.2)} \qquad \frac{S_3 - S_2}{S_2} \leqq \frac{2}{3}$$

which means that IR will never use more than 67% more storage then DS, and this upper bound is attained for $\nu = 1$ and a large NZ. This shows that if no fill-ins are generated with the DS then this is more efficient. This is illustrated in Table 4.14 where we show results with five Harwell test matrices ([24]). These matrices produce no fill-ins and are close to the worst case for IR which is seen to use up to 53% more storage (using formula (6.1)) and 29% more time. In a more realistic 'cheap' problem we shall expect $\nu \geqq 2$ and the extra storage with IR is even smaller and the extra computing time smaller yet.

Keep in mind that with an 'expensive' problem the reduction in storage and computing time is substantial when using IR with a large T and this is really the kind of problems where space and time matters (cf. Tables 4.8 and 4.9).

				computing time		extra time
Matrix	n	NZ	NZ/n^2	Y12M-DS	Y12M-IR	%
SHL 0	663	1687	.004	0.89	1.11	25
SHL 200	663	1726	.004	0.97	1.16	20
SHL 400	663	1712	.004	0.93	1.12	20
STR 0	363	2454	.019	0.68	0.86	26
BT 0	822	3276	.005	1.24	1.60	29

Table 4.14

Results obtained in the runs with 5 Harwell matrices. No fill-ins are produced in the factorization of these matrices.

Three strategies have been proposed for the practical implementation of iterative refinement

A. The classical or English way: The residual vectors r_i in (1.4) are accumulated in extended precision and then rounded to single precision. All other computations are performed in single precision. This strategy is analyzed in [85] and [86], see also [77], and is implemented in Y12M.

B. The revolutionary or Polish way: In a recent paper ([55]) it is shown that under certain conditions the extended precision is not even needed for the residuals. The result is important for computers/compilers which do not have extended precision. We shall not achieve full machine accuracy of the solution, but the solution process is computationally stable, see also [75].

C. The cautious or Scandinavian way: The vectors r_i, d_i, x_i are stored in extended precision and all inner products in (1.4) - (1.6) are accumulated in extended precision. Everything else is performed in single precision. If the length of single and extended precision numbers are n_1 and n_2 digits, respectively and $n_2 \geqq 2 \cdot n_1$, then n_1 digits can be gained compared to strategy A provided the iterative process is convergent. This result was shown in [3], [4] and [6] for an algorithm developed in [10] and our experiments indicate that it holds for Gaussian elimination as well.

The price we have to pay is extra storage for the arrays r_i, d_i, x_i, extra time for each iteration, and a few (usually 3 - 4) extra iterations for the extra accuracy. In addition we get more reliable error estimates and possibly a more robust algorithm which may converge in some cases where strategy A does not.

4.7 Iterative refinement and least squares problems

General methods for linear least squares problems are considered in chapter 5 but when such systems are solved by augmentation the resulting linear systems can be treated with the methods from chapters 1 - 4.

Consider the linear least squares problem

(7.1) $$Ax = b$$

where $A \in \tilde{R}^{m \times n}$ and $b \in \tilde{R}^{m \times 1}$ are given and $x \in \tilde{R}^{n \times 1}$ is to be found such that the Euclidean norm of

(7.2) $$r = b - Ax$$

is as small as possible. The augmented system is

(7.3) $$By = c$$

where

$$(7.4) \qquad B = \begin{Bmatrix} I & A \\ A^T & 0 \end{Bmatrix}, \quad y = \begin{Bmatrix} r \\ x \end{Bmatrix} \quad \text{and} \quad c = \begin{Bmatrix} b \\ 0 \end{Bmatrix}.$$

Many experiments ([96], [103], [107], [109]) have shown that the use of a large drop-tolerance, T, together with iterative refinement is very efficient when solving systems of the form (7.3). In this section we shall present some numerical results from an experiment with 10 testmatrices, $A = F2(1500, n, 125, 6, 100.0)$, $n = 500(100)1400$. Note that by this choice of the parameters m, c and r the augmented matrices have different dimensions $(n + m = 2000(100)2900)$ but the same number of non-zero elements $(NZ = 2(r \cdot m + 110) + m = 19720)$. The DS option was used with $T = 10^{-12}$ and in two runs with the lower triangular matrix, L, kept and removed, respectively.

The IR option was used with $T = 10^{-k}$, $k = 0(1)4$. The right hand side vector was chosen such that the first m components of y were 0 and the last n components 1. In Tables 4.15 - 17 we give the storage needed (as measured by the parameter COUNT), the computing time used and the accuracy obtained, respectively. The experiment was carried out at NEUCC on an IBM 3033 using the FORTH compiler.

Table 4.15 shows clearly how the storage requirements are reduced when L is not kept (DS option) but that even greater reductions are achieved by (retaining L and) using a large drop tolerance - the bigger the better.

Table 4.16 shows that the computing time for the DS option is fairly insensitive to whether L is kept or not, but that IR and a large drop tolerance is a significant improvement. The optimum drop tolerance seems to be $T = 0.1$, although $T = 1$ is sometimes faster.

Table 4.17 shows that the accuracy is reasonably good with the DS-option but considerably better with IR (often close to the machine precision) as long as T is less than 1. $T = 1$ is clearly too big as witnessed by the accuracy and the number of iterations (Table 4.16).

n	Direct solution		Iteratively refined solution				
	L kept	L removed	$T = 10^{-4}$	$T = 10^{-3}$	$T = 10^{-2}$	$T = 10^{-1}$	$T = 10^{0}$
500	255889	111192	101610	79481	64110	55412	32129
600	294191	136040	88077	68227	54746	47634	26915
700	300395	140801	69227	57919	47160	39758	23930
800	270602	136181	66526	51133	44452	37907	23432
900	260079	138383	63872	51146	44535	36541	19771
1000	239894	126535	62712	49882	42650	35036	19782
1100	204261	96729	63175	50419	42972	35963	19812
1200	217605	106094	64716	52070	44533	35888	19910
1300	181777	84174	65606	52073	44971	34946	20585
1400	175433	79005	64180	53263	46794	36833	21529

Table 4.15

The storage needed (as measured by COUNT) when solving (7.3) with A = F2(1500, n, 125, 6, 100.0), n = 500(100)1400.

n	Direct solution		Iteratively refined solution				
	L kept	L removed	$T = 10^{-4}$	$T = 10^{-3}$	$T = 10^{-2}$	$T = 10^{-1}$	$T = 10^{0}$
500	130.15	151.71	18.74 (4)	12.68 (4)	9.49 (5)	8.11 (9)	3.30 (3)
600	178.38	188.47	16.48 (3)	11.12 (4)	8.00 (5)	6.87 (9)	2.57 (3)
700	182.78	165.66	12.10 (3)	9.03 (4)	6.78 (5)	5.63 (9)	2.65 (6)
800	137.86	140.22	11.16 (3)	7.67 (4)	6.29 (5)	5.27 (8)	4.73 (24)
900	127.86	133.30	10.41 (3)	7.49 (4)	6.24 (5)	5.13 (8)	5.15 (37)
1000	108.55	113.93	9.55 (3)	7.06 (4)	6.06 (5)	4.99 (9)	2.89 (14)
1100	75.88	72.44	9.48 (3)	6.99 (4)	5.93 (5)	5.13 (8)	3.10 (15)
1200	82.68	80.25	9.58 (3)	7.20 (4)	6.06 (5)	4.98 (7)	7.28 (50)
1300	57.55	56.45	9.91 (4)	7.07 (4)	6.14 (5)	4.91 (8)	5.45 (33)
1400	51.39	48.62	9.25 (3)	7.17 (4)	6.46 (5)	5.17 (8)	4.24 (21)

Table 4.16

The computing time (in seconds on an IBM 3033) when solving (7.3) with A = F2(1500, n, 125, 6, 100.0), n = 500(100)1400. The number of iterations is given in brackets for the IR solutions.

n	Direct solution		Iteratively refined solution				
	L kept	L removed	$T = 10^{-4}$	$T = 10^{-3}$	$T = 10^{-2}$	$T = 10^{-1}$	$T = 10^{0}$
500	3.5 E-5	2.6 E-5	1.0 E-6	1.0 E-6	1.0 E-6	1.0 E-6	2.3 E-4
600	3.0 E-5	2.6 E-5	1.3 E-6	1.3 E-6	1.3 E-6	1.3 E-6	5.9 E-5
700	2.5 E-5	3.4 E-5	4.3 E-6	4.3 E-6	4.3 E-6	4.3 E-6	3.1 E-4
800	2.0 E-3	3.1 E-3	1.2 E-7	1.2 E-7	1.2 E-7	1.2 E-7	1.4 E-5
900	2.6 E-3	3.2 E-3	6.0 E-8	6.0 E-8	6.0 E-8	6.0 E-8	6.0 E-8
1000	2.1 E-3	8.7 E-4	6.0 E-8	6.0 E-8	6.0 E-8	6.0 E-8	3.5 E-4
1100	4.6 E-4	1.5 E-3	6.0 E-8	6.0 E-8	6.0 E-8	6.0 E-8	7.1 E-4
1200	3.5 E-3	1.3 E-3	6.0 E-8	6.0 E-8	6.0 E-8	6.0 E-8	6.0 E-8
1300	1.3 E-3	8.3 E-4	6.0 E-8	6.0 E-8	6.0 E-8	6.0 E-8	6.0 E-8
1400	1.3 E-3	1.7 E-3	6.0 E-8	6.0 E-8	6.0 E-8	6.0 E-8	6.0 E-8

Table 4.17

The accuracy obtained when solving (7.3) with
A = F2(1500, n, 125, 6, 100.0), n = 500(100)1400.

The stop criterion is a combination of (1.7), (1.8) and (1.9) with MAXIT = 50. When T = 1 divergence is obvious for n = 500 and 600, but also for the other values of n a closer examination of $\|d_q\|$ would indicate that the iteration is too erratic to be trusted.

This is of course only one experiment showing the superiority of IR and a large T. We can produce no proof nor issue any guarantee that this option is always better, but many other experiments have shown it always to perform very efficiently.

We have tried to solve the same problems with MA28. The results for A = F2(1500, 500, 125, 6, 100.0) are summarized in Table 4.18. One hour CPU time was not sufficient to solve (7.3) with n = 600 and we have not tried other values of n.

The failure of MA28 on matrices of the form (7.4) deserves some explanation - and it will transpire that such matrices are especially nasty for MA28 and not typical for its general performance. The

Accuracy	1.7 E-2
Computing time	1466.04
Storage (COUNT)	271356

Table 4.18

Some characteristics obtained in the solution of (7.3) with $A = F2(1500, 500, 125, 6, 100.0)$ using MA28 on an IBM 3033 at NEUCC.

number of non-zero elements in 1480 out of the 1500 rows of $F2(1500, n, c, r, \alpha)$ is r (and is between $r+1$ and $r+10$ in the remaining 20 rows). Therefore the number of non-zero elements in 1480 out of $1500+n$ rows of B is $r+1$ (and greater in the remaining rows if $n < 750$). Similarly for the columns since B is symmetric. The minimum Markowitz cost is thus $r \cdot r$ and is achieved at 1480 diagonal elements which however must all be rejected as pivotal elements because of the stability criterion (cf. (3.1.2) and (3.3.4)). But a total of 1480 rows and 1480 columns must be searched in the first stage of the Gaussian elimination before the GMS realizes that the minimum Markowitz cost for an element that satisfies the stability criterion is greater than or equal to $r \cdot (2r-1)$ (if $n < 750$). In the following many stages the situation is quite similar. The advantage of searching only few rows (as in Y12M) is clear. It should be mentioned that searching a few rows only is now introduced in MA28 (see Harwell Subroutine Library Bulletin, No. 16 (1983), p.2).

4.8 Condition number estimation

The condition number, $\kappa(A) = \|A\| \cdot \|A^{-1}\|$, of the coefficient matrix is rather important, not only for the convergence of the iterative refinement process, but also as a measure of the sensitivity of the solution to (round-off) errors. We shall not compute $\kappa(A)$ according to its definition since this is rather expensive and an accurate value is not needed.

Reliable and inexpensive condition number estimates can be obtained using the algorithms proposed in [13], [14], [35] and [60]. These are all derived for dense matrices (for band matrices see [50]) but they are equally valid for sparse matrices.

A subroutine based on the algorithm described in [35] is included in the package Y12M and can optionally be called to calculate an estimate, COND, of the condition number of the coefficient matrix. Note that the lower triangular matrix, L, should be kept when this subroutine is to be called.

From the discussion in section 4.2 it appears that the drop tolerance, T, should be at most of the order of the reciprocal of COND in order for the iterative refinement process to converge, and a calculation of COND is therefore a good help when trying to find a good value of T.

If COND is of the order of the reciprocal of the machine accuracy then iterative refinement cannot be expected to converge for any T and we must resort to extended precision to solve the problem.

4.9 Robustness and reliability

So far we have mainly been discussing algorithms for sparse matrices. In order to turn an algorithm into a piece of software we must require robustness and reliability and, if we want the software to be used, a certain amount of efficiency.

By robustness we mean that

(a) the code should only give up if a problem is really hard, and

(b) if the code quits, it should give good information on whether failure was due to

- (b1) ill-condition of the problem,
- (b2) instability of the elimination,
- (b3) insufficient storage,
- (b4) divergence of the iterative process,
- (b5) or something else.

By reliability we mean that

(c) the code should never give a bad answer pretending it is good, and

(d) the code should provide error estimates.

To aid the user maintaining efficiency in the computations the code should also in case of success give feed-back on important details such as

(e1) how much storage was actually used,
(e2) how many iterations, and
(e3) the proportion of time for factorization and iterations.

The user must in turn be provided with a number of handles to turn in response to the information from (b) and (e) such as

(f1) the stability factor (u),
(f2) the drop tolerance (T),
(f3) the number of rows to search (p)
(f4) the sizes of the major arrays (NN, NN1),
(f5) etc.

But the user should not be burdened unnecessarily by all these parameters so we should have

(g) default values for the parameters.

In order to take advantage of special situations the code should also have

(h) options for special matrices or problems.

It is our experience that Gaussian elimination with IGMS combined with iterative refinement and a large drop tolerance provides a good basis for a robust, reliable and efficient sparse matrix code. The points (a), (c), (d) and (f) are taken care of by our discussion in the previous sections. The points (b), (e) and (h) should be kept in mind when implementing the algorithm and as for the default values we can recommend

$$u \in [4,10]$$

$$T \in [0.01,\ 0.001],$$

$$p \in \{2,3\},$$

$$NN \simeq 3 \cdot NZ,$$

$$NN1 \simeq 0.6 \cdot NN.$$

These recommendations must of course be taken with a grain of salt as they are very dependent on the problem and the option. This is also why we recommend point (b) and (e) such that proper action can be taken when a related problem is to be solved.

Keep in mind the classification of problems from section 2.7 to which we can add, that with IR category (1) turns into (2) and category (3) turns into (4). However, we might keep the option of turning IR off and compute the direct solution without retaining the matrix L (see section 2.6).

Ill-condition of a problem (b1) will often, but not always, be detected by very small pivots. On the other hand, small pivots may be caused by bad scaling. Instability of the elimination (b2) can be detected by monitoring b_k (see (3.1.4)) and can be counteracted by reducing the stability factor u. Insufficient storage (b3) should be reported explicitly with indication of which array(s) need expansion and at what stage of the elimination.

The iterative refinement process (b4) may converge slowly or diverge and both cases are identified by the value of RELEST and the number of iterations. Non-convergence may be caused by a too large T or an ill-conditioned coefficient matrix. For problems of categories (3) and (4) the 'variable-T-strategy' mentioned in section 4.5 might be very useful for finding an optimal T.

We have already (in section 3.4) mentioned special classes of matrices for which special pivoting strategies can be applied with success. We recommend that a sparse matrix code have options to treat such special cases efficiently, as well as an option to turn IR off.

4.10 Concluding remarks on IR and T

The process of computing a crude but cheap decomposition of matrix A by employing a large drop tolerance can be considered as a "preconditioning" of matrix A. The LU factorization obtained in this way is sometimes referred to as incomplete or partial. The process of preconditioning by incomplete factorization has successfully been used for symmetric positive definite matrices in conjunction with some iterative method such as SOR or conjugate gradient (see [2], [27], [33], [34], [56], [58], [63], [81]). These iterative methods can not be used with general matrices, but (as in Y12M) iterative refinement can be used instead. Viewed in this way the title of this book might as well have been: Incomplete LU-decomposition as preconditioning for iterative refinement.

Our experiments have shown that this approach is often very efficient (see [96], [107], [109]). But it should be emphasized that there exist problems for which the direct solution is competitive and therefore the iterative refinement process should only be an option in a package for sparse linear systems. Note too, that if matrix A is very ill-conditioned, then the IR process may not converge even with $T = 0$. The DS option with double precision computation "might give an answer of acceptable (but unknown) accuracy" in this case (see Golub and Wilkinson [49, p. 148]). It should be mentioned here that by the use of some machine-dependent facilities (paging, multibanking, etc.) one can modify the iterative refinement option so that it will never use more storage than the direct solution option. The price to be paid for this is a modest increase of the computing time. A modified version of the IR option of Y12M using multibanking has been developed for the UNIVAC 1100/82 at RECKU ([82]).

Chapter 5: Other Direct Methods

5.1 Linear least squares problems

Let m and n be integers, $b \in \tilde{C}^{m\times 1}$ a vector and $A \in \tilde{C}^{m\times n}$ a matrix. By A^H we denote the conjugate transpose of A.

Definition 5.1 The unique matrix $A^+ \in \tilde{C}^{n\times m}$ satisfying the conditions

(1.1) $$A^+ A A^+ = A^+,$$

(1.2) $$A A^+ A = A,$$

(1.3) $$(A^+ A)^H = A^+ A,$$

(1.4) $$(A A^+)^H = A A^+$$

is called the Moore-Penrose generalized inverse or pseudo-inverse of matrix A. ∎

Remark 5.2 Moore [61] was probably the first who introduced the generalized inverse of a matrix. The conditions (1.1) - (1.4) were formulated considerably later by Penrose ([65]). ∎

Remark 5.3 If $m = n$ and $\operatorname{rank}(A) = n$ then A^{-1} satisfies (1.1) - (1.4). This fact justifies the term generalized inverse for A^+. ∎

Definition 5.4 The linear least squares problem is the problem of finding a vector $x \in \tilde{C}^{n\times 1}$ which minimizes the Euclidean norm of

(1.5) $$r = b - Ax, \qquad (r \in \tilde{C}^{m\times 1}).$$

x is called a least squares solution. ∎

<u>Theorem 5.5</u> All solutions of (1.5) are given by

$$x = A^{+}b + (I - A^{+}A)z, \tag{1.6}$$

where $z \in \tilde{C}^{n\times 1}$ is an arbitrary vector.

<u>Proof</u> See [79]. ∎

<u>Corollary 5.6</u> The least squares solution of (1.5) which has minimal Euclidean norm is unique and equal to $A^{+}b$. ∎

<u>Corollary 5.7</u> If $\operatorname{rank}(A) = n$ then the least squares solution of (1.5) is unique and equal to $A^{+}b$. ∎

In this chapter we shall consider direct methods for sparse, real least squares problems, where A has full column rank, i.e. we shall assume

$$A \in \tilde{R}^{m\times n}, \quad b \in \tilde{R}^{m\times 1}, \tag{1.7}$$

$$\operatorname{rank}(A) = n, \tag{1.8}$$

$$A \text{ is large and sparse.} \tag{1.9}$$

<u>Remark 5.8</u> It follows from (1.7) and (1.8) that

$$m \geq n \quad \text{and} \quad x \in \tilde{R}^{n\times 1}. \tag{1.10}$$ ∎

<u>Remark 5.9</u> The condition (1.8) is essential for the methods we are about to discuss. If $\operatorname{rank}(A) < n$ then other methods such as the Singular Value Decomposition (see [47], [48], [77]) should be used. ∎

<u>Lemma 5.10</u> If (1.7) and (1.8) are satisfied then

$$A^{+} = (A^{T}A)^{-1}A^{T}. \tag{1.11}$$ ∎

Remark 5.11 If (1.7) and (1.8) are satisfied then the linear least squares problem (1.5) can be reformulated as:

Solve the system $Ax = b - r$
under the condition $A^T r = 0$.

It is thus equivalent to the $(m+n) \times (m+n)$ linear system

$$(1.12) \qquad \left[\begin{array}{c|c} I & A \\ \hline A^T & 0 \end{array}\right] \cdot \left[\begin{array}{c} r \\ \hline x \end{array}\right] = \left[\begin{array}{c} b \\ \hline 0 \end{array}\right]$$

(cf. section 4.7). ∎

5.2 The general k-stage direct method

Taking the unavoidable computational errors into account we shall replace the linear least squares problem with a weaker one.

Problems 5.12 Find an approximation $\bar{x} \in \mathbb{R}^{n \times 1}$ to the least squares solution $x = A^+ b$.

In this section we shall introduce a general computational scheme which includes many of the so-called direct methods for the linear least squares problem.

Assume that it is possible to replace Problem 5.12 with the following.

Problem 5.13 Find a vector $y \in \mathbb{R}^{\bar{q} \times 1}$ such that

$$(2.1) \qquad y = B_1^+ c,$$

where

$$(2.2) \qquad \bar{p} \geq \bar{q}, \quad \bar{p} \in \mathbb{N}, \quad \bar{q} \in \mathbb{N},$$

$$(2.3) \qquad B_1 \in \mathbb{R}^{\bar{p} \times \bar{q}},$$

(2.4) $\mathrm{rank}(B_1) = \bar{q}$,

(2.5) B_1 and c can be computed from A and b,

(2.6) there is a simple relationship between x and y. ∎

An approximation y_1 to y can be obtained through the following two computational steps:

Step 1 - Generalized decomposition.

Compute

$$\bar{B}_i = P_i B_i Q_i + E_i, \qquad i = 1(1)k, \qquad k \in \tilde{N}, \tag{2.7}$$

where P_i and Q_i are permutation matrices and E_i are perturbation-(error-) matrices. $\bar{B}_i$ is assumed to be decomposed

$$\bar{B}_i = C_i \bar{C}_i D_i, \qquad i = 1(1)k, \tag{2.8}$$

and if $k > 1$ we have

$$B_{i+1} = C_i^T C_i \bar{C}_i, \qquad i = 1(1)k-1. \tag{2.9}$$

We demand that D_i are such that $D_i^+ z$, $i = 1(1)k$, can be easily computed for any vector z, and furthermore that the decomposition of $\bar{B}_k$ is such that $\bar{B}_k^+ z$ can be easily computed.

Apart from this we put no restrictions on the matrices $\bar{B}_i$, C_i, $\bar{C}_i$, D_i except that the dimensions match such that all multiplications can be carried out.

Step 2 - Generalized substitution.

Compute

$$y_1 = \left(\prod_{i=1}^{k-1} Q_i D_i^+\right) Q_k \bar{B}_k^+ P_k \left(\prod_{i=1}^{k-1} P_i^T C_i\right)^T c = Hc. \tag{2.10}$$

Remark 5.14 In (2.10) and following expressions we use

$$(2.11) \qquad \prod_{i=j}^{k} A_i = A_j \cdot A_{j+1} \cdots A_k \qquad \text{when} \quad k \geq j$$

and

$$(2.12) \qquad \prod_{i=j}^{k} A_i = I \qquad \text{when} \quad k < j. \qquad \blacksquare$$

If y_1 is an approximation to y in (2.1) then we can use y_1 and the relationship (2.6) to obtain an approximation $\bar{x}$ to the least squares solution x. Therefore we must prove that y_1 will be a good approximation to y when the perturbation matrices E_i, $i = 1(1)k$, are "small" in some sense.

With $H \in \tilde{R}^{\bar{q} \times \bar{p}}$ from (2.10) define

$$(2.13) \qquad F = I - HB_1 \qquad (F, I \in \tilde{R}^{q \times q}).$$

Then we have the following theorem ([93]).

Theorem 5.15 Assume that $\bar{B}_k$ and D_i, $i = 1(1)k-1$, have full column rank. Then

$$(2.14) \qquad F = \sum_{j=1}^{k} H_j$$

where

$$(2.15) \qquad H_j = M_j P_j^T E_j Q_j^T \left(\prod_{i=1}^{j-1} Q_i D_i^T \right)^T$$

and

$$(2.16) \qquad M_j = \left(\prod_{i=1}^{k-1} Q_i D_i^+ \right) Q_k \bar{B}_k^+ P_k \left(\prod_{i=j}^{k-1} P_i^T C_i \right)^T .$$

Proof If $k = 1$ then

$$F = I - HB_1 = I - Q_1 \bar{B}_1^+ P_1 B_1 = Q_1 (I - \bar{B}_1^+ P_1 B_1 Q_1) Q_1^T$$
$$= Q_1 \bar{B}_1^+ E_1 Q_1^T = Q_1 \bar{B}_1^+ P_1 P_1^T E_1 Q_1^T = M_1 P_1^T E_1 Q_1^T = H_1$$

If $k \geqq 2$ we get using (2.13), (2.10), (2.16), (2.11), (2.7), (2.8), (2.9), (2.15) and (2.12)

$$\begin{aligned} F &= I - M_1 B_1 = I - M_2 C_1^T P_1 B_1 \\ &= I - M_2 C_1^T (\bar{B}_1 - E_1) Q_1^T \\ &= I - M_2 C_1^T C_1 \bar{C}_1 D_1 Q_1^T + M_2 C_1^T P_1 P_1^T E_1 Q_1^T \qquad (2.17) \\ &= I - M_2 B_2 \left(\prod_{i=1}^{1} Q_1 D_1^T \right)^T + H_1 . \end{aligned}$$

This is the beginning of an induction argument where the induction step is $(2 \leqq j \leqq k-1)$:

$$\begin{aligned} M_j B_j \left(\prod_{i=1}^{j-1} Q_i D_i^T \right)^T &= M_{j+1} C_j^T P_j B_j \left(\prod_{i=1}^{j-1} Q_i D_i^T \right)^T \\ &= M_{j+1} C_j^T \left(\bar{B}_j - E_j \right) Q_j^T \left(\prod_{i=1}^{j-1} Q_i D_i^T \right)^T \qquad (2.18) \\ &= M_{j+1} B_{j+1} \left(\prod_{i=1}^{j} Q_i D_i^T \right)^T - H_j . \end{aligned}$$

In the final step we use (2.12), (2.7), (2.16) and (2.15) together with the assumption that $\bar{B}_k$ and D_i have full column rank:

$$\begin{aligned} F &= I - M_k B_k \left(\prod_{i=1}^{k-1} Q_i D_i^T \right)^T + \sum_{j=1}^{k-1} H_j \\ &= I - \left(\prod_{i=1}^{k-1} Q_i D_i^+ \right) Q_k \bar{B}_k^+ P_k B_k \left(\prod_{i=1}^{k-1} Q_i D_i^T \right)^T + \sum_{j=1}^{k-1} H_j \qquad (2.19) \\ &= I - \left(\prod_{i=1}^{k-1} Q_i D_i^+ \right) Q_k \bar{B}_k^+ \left(\bar{B}_k - E_k \right) Q_k^T \left(\prod_{i=1}^{k-1} Q_i D_i^T \right)^T + \sum_{j=1}^{k-1} H_j \\ &= H_k + \sum_{j=1}^{k-1} H_j = \sum_{j=1}^{k} H_j . \end{aligned}$$

∎

Corollary 5.16 If the decomposition is performed with no errors, i.e. $E_i = 0$, $i = 1(1)k$, then $H_i = 0$, $i = 1(1)k$, and $H = B_1^+$.

If moreover the substitution is performed without rounding errors then $y_1 = y$. ∎

Definition 5.17 The computational scheme given by Step 1 and Step 2 is called a general k-stage direct method or k-stage computational scheme for solving Problem 5.13. ∎

5.3 Special cases of the general method

We now give six examples of well-known and commonly used direct methods which can be viewed as special cases of the general k-stage computational scheme. Most of the methods are 1-stage methods and for $k = 1$ the general method reduces to

$$\bar{B}_1 = P_1 B_1 Q_1 + E_1, \tag{3.1}$$

$$\bar{B}_1 = C_1 \bar{C}_1 D_1, \tag{3.2}$$

$$y = Hc = Q_1 \bar{B}_1^+ P_1 c. \tag{3.3}$$

We must therefore specify C_1, $\bar{C}_1$ and D_1 and verify that $\bar{B}_1^+ z$ is easily computed for arbitrary z.

Example 5.18 If $m = n$ the classical Gaussian elimination is obtained from the general scheme by setting $k = 1$ and

$$B_1 = A, \quad c = b, \quad y = x; \tag{3.4a}$$

$$C_1 = L_g, \quad \bar{C}_1 = I, \quad D_1 = U_g; \tag{3.4b}$$

$$\bar{x} = y_1. \tag{3.4c}$$

(3.4a) is the transformation from Problem 5.12 to Problem 5.13, (3.4b) specifies the method, and (3.4c) the relationship between y_1 and $\bar{x}$. L_g and U_g are triangular factors of A as computed by Gaussian elimination. ∎

Example 5.19 Let $m > n$ and assume that the normal equations are solved by some symmetric version of Gaussian elimination. This scheme is obtained by setting $k = 1$ and

(3.5a) $$B_1 = A^T A, \quad c = A^T b, \quad y = x;$$

(3.5b) $$C_1 = L_c, \quad \bar{C}_1 = D_c, \quad D_1 = L_c^T;$$

(3.5c) $$\bar{x} = y_1.$$

Here L_c and D_c are the computed factors in the $L_c D_c L_c^T$-factorization of the positive definite matrix $A^T A$. ∎

Denote by

(3.6) $$\sigma_1 \geq \sigma_2 \geq \dots \geq \sigma_n > 0$$

the singular values of matrix A (the square roots of the eigenvalues of $A^T A$). The spectral condition number of matrix A is

(3.7) $$\kappa(A) = \|A\|_2 \cdot \|A^+\|_2 = \sigma_1/\sigma_n.$$

It is seen that

(3.8) $$\kappa(B_1) = (\kappa(A))^2,$$

and this is one reason why the normal equations cannot be generally recommended for the solution of linear least squares problems. Furthermore, if A is large and sparse and not too well-conditioned then (3.8) may restrict the choice of drop tolerance severely - and $A^T A$ may not be very sparse ([7], [8]).

Example 5.20 Let $m > n$ and assume that the augmented linear system is solved by Gaussian elimination. This method is obtained with $k = 1$ and (cf. (1.12))

(3.9a) $$B_1 = \left[\begin{array}{c|c} \alpha I & A \\ \hline A^T & 0 \end{array}\right], \quad c = \left[\begin{array}{c} b \\ \hline 0 \end{array}\right], \quad y = \left[\begin{array}{c} \alpha^{-1} r \\ \hline x \end{array}\right];$$

(3.9b) $\quad C_1 = L_a, \quad \bar{C}_1 = I, \quad D_1 = U_a;$

(3.9c) $\quad \bar{x}$ = the last n coordinates of y_1.

B_1 is the so-called augmented matrix and L_a and U_a are its triangular factors as computed by Gaussian elimination.

Björck ([3], [4]) has shown that

(3.10) $\quad \kappa(B_1) \approx \sqrt{2} \cdot \kappa(A) \qquad$ for $\alpha = \sigma_n/\sqrt{2}$

and

(3.11) $\quad \kappa(B_1) \gtrapprox (\kappa(A))^2 \qquad$ for $\alpha \gtrapprox \sigma_1/\sqrt{2}$

Duff & Reid ([23]) have reported good results with $\alpha = 1$ but as (3.11) indicates, α must be chosen carefully. Note that $\kappa(B_1)$ increases again for $\alpha < \sigma_n/\sqrt{2}$. (See also [7]). ∎

<u>Example 5.21</u> The Peters-Wilkinson method ([66]) can be obtained from the general scheme by choosing $k = 2$ and

(3.12a) $\quad B_1 = A, \quad c = b, \quad y = x;$

(3.12b) $\quad C_1 = L_p, \quad \bar{C}_1 = I, \quad D_1 = U_p;$

(3.12c) $\quad C_2 = \bar{L}_p, \quad \bar{C}_2 = \bar{D}_p, \quad D_2 = \bar{L}_p^T;$

(3.12d) $\quad \bar{x} = y_1.$

Here L_p is an $m \times n$ unit lower trapezoidal matrix, U_p is an $n \times n$ upper triangular matrix and $L_p U_p = A + E_1$. $\bar{L}_p \bar{D}_p \bar{L}_p^T$ is the computed decomposition of the symmetrix matrix $B_2 = L_p^T L_p$.

One can expect that the computations in the second stage will be about as accurate as those in the first stage if $\kappa(L_p) \approx \sqrt{\kappa(A)}$. We cannot prove any such relation, but heuristic considerations indicate that ill-conditioning of A will normally be reflected in U_p, and that L_p is often well-conditioned ([14], [66]).

Numerical evidence shows that the method is often numerically stable (see e.g. [23]). ∎

Example 5.22 An orthogonal decomposition of A can be obtained from the general scheme with $k = 1$ and

(3.13a) $$B_1 = A, \quad c = b, \quad y = x;$$

(3.13b) $$C_1 = R, \quad \bar{C}_1 = D, \quad D_1 = S;$$

(3.13c) $$x = y_1;$$

Here $R \in \tilde{R}^{m\times n}$ is orthogonal, i.e. $R^T R = I_{n\times n}$, D is a diagonal $n \times n$ matrix, S is an upper triangular $n \times n$ matrix and RDS is the orthogonal decomposition of A. If this is computed by Householder's or Givens' method then $D = I$. ∎

Example 5.23 Another version of the orthogonal decomposition is derived by setting $k = 2$, $P_1 = P_2 = Q_1 = Q_2 = I$ and using the decomposition $PAQ + E = RDS$ from Example 5.22. We define

(3.14a) $$B_1 = A^T A, \quad c = A^T b, \quad y = x;$$

(3.14b) $$C_1 = I, \quad \bar{C}_1 = A^T P^T R, \quad D_1 = DSQ^T;$$

(3.14c) $$C_2 = Q, \quad \bar{C}_2 = S^T, \quad D_2 = D;$$

(3.14d) $$\bar{x} = y_1.$$

In this case

(3.15) $$\begin{aligned} y_1 &= Hc = Q_1 D_1^+ Q_2 \bar{B}_2^+ P_2 C_1^T P_1 c = (DSQ^T)^{-1} (QS^T D)^{-1} c \\ &= QS^{-1} D^{-2} (S^T)^{-1} Q^T A^T b. \end{aligned}$$

Note that all matrices needed in the computation of y_1 are to be computed in the first stage and that matrix R does not participate in the actual computations and therefore need not be stored. If we are using a dense matrix technique we have room for the information needed to retrieve R below the diagonal of A

(see [78]) and we might as well store it, but if A is large and sparse and a sparse matrix technique is used we can save a considerable amount of space since R is often much less sparse than A. This fact is emphasized in [5], [7] and exploited in a code developed by Zlatev and Nielsen [102] (see section 5.7). Note however that both A and S should be stored. For the perturbation matrices E_1 and E_2 we have the following expressions

$$\begin{aligned} E_1 &= \bar{B}_1 - P_1 B_1 Q_1 = (A^T P^T R)(DSQ^T) - A^T A \\ &= A^T P^T (RDS - PAQ) Q^T = A^T P^T E Q^T, \end{aligned} \tag{3.16}$$

and

$$\begin{aligned} E_2 &= \bar{B}_2 - P_2 B_2 Q_2 = QS^T D - A^T P^T R \\ &= Q \cdot (S^T D^T R^T - Q^T A^T P^T) R = QE^T R. \end{aligned} \tag{3.17}$$

∎

5.4 Generalized iterative refinement

We shall now focus our attention on linear least squares problems where the coefficient matrix, A, is large and sparse. In order to solve such problems efficiently we must minimize the storage and computation time for the solution process. To achieve this we shall employ a sparse matrix technique, select a proper pivoting strategy, choose a reasonable stability factor, and use a large drop tolerance (see e.g. [12], [68] and [89]).

These attempts to exploit and preserve the sparsity - and in particular the last point - will often be accompanied by a loss of accuracy which we should somehow try to regain. This can usually be done by adding iterative refinement to the general k-stage scheme as we did in Chapter 4 for Gaussian elimination. We shall therefore add the following step to the two computational steps of section 5.2.

<u>Step 3</u> - Generalized iterative refinement.

(4.1) $$r_i = c - B_1 y_i \quad , \qquad i = 1(1)q-1;$$

(4.2) $$d_i = H r_i \quad , \qquad i = 1(1)q-1;$$

(4.3) $$y_{i+1} = y_i + d_i \quad , \qquad i = 1(1)q-1.$$

Some stop-criteria (see [6], [77]) must be used to terminate the iterative process, and y_q will be accepted as an approximation to y. Finally $\bar{x}$ must be found from y_q using the relationship between x and y.

For the moment we shall assume that (2.10) and (4.1) - (4.3) can be performed without rounding errors. Define

(4.4) $$s = c - B_1 y \quad , \qquad s \in \tilde{R}^{\bar{p} \times 1}.$$

From (2.1) and (1.11) it follows that

(4.5) $$B_1^T s = 0.$$

We shall need the following theorems in the discussion of the convergence of the iterative process (4.1) - (4.3).

<u>Theorem 5.24</u> If $\{y_i\}$ is the sequence of vectors calculated by (2.10) and (4.3), then

(4.6) $$y_i = y + F^{i-j}(y_j - y) + \left(\sum_{\nu=0}^{i-j-1} F^{\nu} \right) Hs$$

for any $j < i$.

<u>Proof</u> It follows from (4.1) - (4.4) and (2.13) that

$$y_i - y = y_{i-1} + d_{i-1} - y$$

(4.7) $$= y_{i-1} + H \cdot (B_1(y - y_{i-1}) + s) - y$$

$$= F(y_{i-1} - y) + Hs,$$

and (4.6) follows easily. ∎

<u>Theorem 5.25</u> If $\{d_i\}$ is the sequence of vectors calculated by (4.2), then

$$d_i = F^{i-j} d_j \tag{4.8}$$

for any $j \leqq i$.

<u>Proof</u> The assertion follows immediately from

$$\begin{aligned} d_i &= d_{i-1} + (d_i - d_{i-1}) \\ &= d_{i-1} + H B_1 (y_{i-1} - y_i) = F d_{i-1}. \end{aligned} \tag{4.9}$$

■

<u>Theorem 5.26</u> If $\rho(F) < 1$, then

$$y = y_k + \sum_{i=k}^{\infty} d_i - (H B_1)^{-1} Hs \tag{4.10}$$

for any fixed positive integer k.

<u>Proof</u> $\rho(F)$ denotes the spectral radius of F and $\rho(F) < 1$ implies

$$\lim_{j\to\infty} F^j = 0 \quad \text{and} \quad \sum_{j=0}^{\infty} F^j = (I - F)^{-1}. \tag{4.11}$$

Therefore we have from (4.6) and (2.13)

$$\lim_{i\to\infty} y_i = y + \left(\sum_{j=0}^{\infty} F^j \right) H s = y + (H B_1)^{-1} H s. \tag{4.12}$$

From (4.3) we find

$$\lim_{i\to\infty} y_i = y_k + \sum_{i=k}^{\infty} d_i \tag{4.13}$$

and (4.10) follows. ■

Corollary 5.27 If $\rho(F) < 1$ then the iterative process (4.1) - (4.3) is convergent to the true solution of (2.1) if one of the following three conditions is satisfied:

(4.14) $$s = 0 \quad \text{or}$$

(4.15) $$H = B_1^+ \quad \text{or}$$

(4.16) $$H = \bar{H} B_1^T$$

where $\bar{H} \in \mathbb{R}^{\bar{p} \times \bar{p}}$ is arbitrary. ∎

Corollary 5.28 The iterative process (4.1) - (4.3) is convergent if $\|F\| < 1$, where $\|\cdot\|$ denotes any matrix norm induced from the vector norm chosen. ∎

Remark 5.29 The condition (4.16) must be characterized as purely theoretical. ∎

The condition (4.14) is satisfied in Examples 5.18 - 5.20 and 5.23 of the preceding section, and in all examples H is an approximation to B_1^+. This means that the iterative process (4.1) - (4.3) is convergent to the true solution of (2.1) provided the computations in (2.10) and (4.1) - (4.3) are performed without errors. Experimental evidence shows, however, that even with the presence of rounding errors good results can be obtained, the reason being that the amount of computation in (2.10) and (4.1) - (4.3) is fairly restricted and the accumulated rounding errors therefore rather small.

Assume that all matrices B_i, $i = 1(1)k$ are well scaled, that b_i is the magnitude of the non-zero elements of B_i and that we choose a drop tolerance T_i at stage i of the k-stage computational scheme. Assume also that some version of Gaussian elimination or some orthogonal decomposition is used for the factorization (2.8).

Then

(4.17) $$\|E_i\| \leq f_i(m,n) \cdot \bar{\varepsilon}_i \cdot g_i(A), \qquad \bar{\varepsilon}_i = \max(\varepsilon, T_i/b_i),$$

where $f_i(m,n)$ is a function of m, n and the factorization method, ε is the machine accuracy, and $g_i(A)$ is some function of $\|A\|$.

Let

$$(4.18) \qquad f(m,n) = \max_{1\leq i\leq k} \{f_i(m,n)\},$$

$$(4.19) \qquad g(A) = \max_{1\leq i\leq k} \{g_i(A)\},$$

$$(4.20) \qquad \bar{g}(A) = \max_{1\leq i\leq k} \left\{ \| M_i P_i^T \| \cdot \| Q_i^T \left(\prod_{j=1}^{i-1} Q_j D_j^T \right)^T \| \right\}$$

(cf. (2.15) and (2.16)).

Then

$$(4.21) \qquad \|F\| \leq k \cdot f(m,n) \cdot \bar{\varepsilon} \cdot g(A) \cdot \bar{g}(A), \qquad \bar{\varepsilon} = \max_{1\leq i\leq k} \{\bar{\varepsilon}_i\}.$$

When $k = 1$ and $T_1 = 0$ $g(A) \cdot \bar{g}(A)$ can often be expressed by the condition number of A. If the spectral condition number

$$(4.22) \qquad \kappa_2(A) = \|A\|_2 \cdot \|A^+\|_2$$

is used in connection with the methods from Example 5.19 and Example 5.23 then $\kappa_2(A)$ can be replaced by

$$(4.23) \qquad \kappa_2'(A) = \inf_{D>0}\{\kappa_2(AD)\},$$

where D is a diagonal matrix with positive elements ([6], p. 163).

The practical value of the bound (4.21) is rather small since the theoretical values for $f(m,n)$ are usually very crude and give severe overestimates for $\|F\|$. But (4.21) and (4.17) indicate an important relationship between the condition number ((4.22) or 4.23)) and the drop tolerance showing once again that if κ is large T must be chosen smaller.

For matrices of class $F2(m,n,c,r,\alpha)$ we can vary the condition number by varying α. The interplay between T and α on such a matrix is shown in Table 5.1 where $\max \alpha$ indicates the largest power of 2 which allows a successful solution. Gaussian elimination is performed with an improved version of the code SIRSM ([99], [100]), the orthogonal transformations are performed with the code LLSS01 ([101]), and COND is an estimate of the condition number found by a FORTRAN subroutine given by [35].

Drop tolerance T	Gaussian elimination			Orthogonal transformations		
	max α	COND	$\|\|x-\bar{x}\|\|_\infty$	max α	COND	$\|\|x-\bar{x}\|\|_\infty$
0	2^{24}	4.43E+14	8.46E-14	2^{15}	1.69E+12	7.09E-7
10^{-4}	2^{16}	6.75E+12	2.79E-11	2^{9}	4.02E+8	7.54E-10
10^{-3}	2^{13}	1.01E+11	6.51E-12	2^{7}	2.35E+7	4.66E-10
10^{-2}	2^{10}	1.69E+9	3.38E-12	2^{4}	2.43E+5	5.17E-12

Table 5.1

Solution with matrices $F2(22, 22, 11, 2, \alpha)$ showing the maximum value of α allowing a successful solution with a given value of T.

5.5 Orthogonal transformations

We shall in section 5.7 take a closer look at the implementation details of a 2-stage scheme based on orthogonal transformations but first we shall discuss the orthogonal RDS decomposition of an $m \times n$ martrix A.

Two approaches have been very popular, in text-books and in practical use: the Givens method ([45], [46]) based on plane rotations and the Householder method ([54]) based on elementary reflectors.

The computational cost of Givens' and Householder's methods, measured by the number of multiplications and square roots for dense matrices is given in Table 5.2 and these figures indicate why Householder's method has been the more popular one since 1959.

Method	Multiplications	Square roots
Givens	$\frac{4}{3}mn^2$	$m \cdot n$
Householder	$\frac{2}{3}mn^2$	n

Table 5.2

Computational costs for dense matrices

Recently the situation has changed due to results of [38], [39], [40], and [53], which have brought the computational cost of Givens' method down to about the same as for Householder, and as Givens' method is favourable with sparse matrices we shall discuss it in more detail.

The orthogonal reduction is performed in n major steps, each one transforming all elements below the diagonal in a certain column to 0. Each major step consists of several minor steps - the plane rotations - each one transforming one element to 0. If there are s_k such elements then the k-th major step will consist of s_k minor steps.

In the ordinary Givens method a minor step consists of the following multiplications

$$(5.1) \quad \begin{bmatrix} \gamma & \sigma \\ -\sigma & \gamma \end{bmatrix} \cdot \begin{bmatrix} d_i & 0 \\ 0 & d_j \end{bmatrix} \cdot \begin{bmatrix} a_{i,k} & a_{i,k+1} & \cdots & a_{i,n} \\ a_{j,k} & a_{j,k+1} & \cdots & a_{j,n} \end{bmatrix}$$

where one of the elements $a_{i,k}$, $a_{j,k}$ shall be transformed to 0 and $d_i = d_j = 1$. Actually the first two matrices are $m \times m$ matrices, but with 1's in the diagonal in all the other rows, so we show only the elements that take part in the computations.

It is clear that two multiplications are needed for each $a_{\cdot,\cdot}$ that take part in the transformation (5.1). To avoid too much work Gentleman has therefore suggested the following refactorization of the first two matrices

$$(5.2)\qquad \begin{bmatrix} \gamma & \sigma \\ -\sigma & \gamma \end{bmatrix} \cdot \begin{bmatrix} d_i & 0 \\ 0 & d_j \end{bmatrix} = \begin{bmatrix} d_i\gamma & 0 \\ 0 & d_j\gamma \end{bmatrix} \cdot \begin{bmatrix} 1 & \alpha \\ \beta & 1 \end{bmatrix}$$

and we only perform the multiplications by the last matrix in (5.2). It is readily seen that only one multiplication (and one addition) is needed for each $a_{\cdot,\cdot}$ with this scheme.

We still have to decide which one of a_{ik} and a_{jk} to transform to 0 and give formulas for α, β and γ:

If

$$(5.3)\qquad d_i^2\, a_{ik}^2 \geqq d_j^2\, a_{jk}^2$$

then $a_{jk} := 0$ with

$$(5.4)\qquad \beta = -\, a_{jk}/a_{ik}, \qquad \alpha = -\,\beta \cdot d_j^2/d_i^2,$$

else $a_{ik} := 0$ with

$$(5.5)\qquad \alpha = -\, a_{ik}/a_{jk}, \qquad \beta = -\,\alpha \cdot d_i^2/d_j^2.$$

In both cases

$$(5.6)\qquad \gamma^2 = \frac{1}{1-\alpha\beta}.$$

The above formulae show that we can avoid square roots - which were necessary in the determination of γ and σ with ordinary Givens - and some multiplications by storing d_i^2 rather than d_i. This is why the name square-root-free Givens has been attached to this method. Furthermore we shall see in a short while that we shall use the matrix D^2 rather than D in our computations (cf. (3.15)).

The d_i are initialized by

$$(5.7) \qquad d_i^2 = 1, \qquad i = 1(1)m,$$

unless the problem is weighted in which case the squares of the weights are used. In each minor step two d_i-s are updated:

$$(5.8) \qquad d_i^2 := d_i^2 \cdot \gamma^2, \quad d_j^2 := d_j^2 \cdot \gamma^2$$

(cf. (5.2)) with γ^2 given in (5.6).

From (5.3) - (5.6) it follows that $\frac{1}{2} \leqq \gamma^2 < 1$ such that elements of D^2 decrease, but not too fast. If the problem is very large underflows may occur, however, and it might be advisable to check the magnitudes of the d_i^2 and rescale the problem if necessary.

Consider again the values in Table 5.2. With the above modifications the Gentleman-Givens method requires roughly the same amount of computational work as Householder's method, but there is still no particular reason to prefer Gentleman-Givens to Householder for dense matrices (note e.g. the underflow problem with the d_i). It should be mentioned that a trapezoidal-triangular LU-decomposition only requires $\frac{1}{3} mn^2$ arithmetic operations but that orthogonal methods usually are preferred since they are believed to be more stable.

When dealing with sparse matrices the preservation of sparsity is an important issue and this swings preference away from Householder.

If $s_k + 1$ rows participate in the computations during major step k of the Householder decomposition then each of the transformed rows is a linear combination of all $s_k + 1$ rows and will therefore have the sparsity pattern of the union of the $s_k + 1$ rows (neglecting cancellations).

For the Givens decomposition the sparsity pattern of the two rows involved in a minor step will be the sparsity pattern of the union of the two rows, and if one row does not take part in any other minor steps (within the major step in question) it will receive no more fill-ins.

For completeness we note that in the trapezoidal-triangular decomposition no fill-ins appear in the pivotal row, and it is thus the best method w.r.t. preserving sparsity.

We illustrate the appearance of fill-ins by a simple example in Fig. 5.3 - 5.6 where we show the original (square) matrix and the matrix after the first major step of Householder, Givens, and Gaussian elimination respectively.

```
x x     x
x   x   x
x     x   x
  x x
  x   x   x
  x x   x
```

<u>Fig. 5.3</u>

The original matrix

```
x x ⊠ ⊠ x ⊠
o ⊠ x ⊠ x ⊠
o ⊠ ⊠ x ⊠ x
  x x
  x   x   x
  x x   x
```

<u>Fig. 5.4</u>

Householder's method

First major step gives 9 fill-ins

```
x x ⊠ ⊠ x ⊠
o ⊠ x   x
o ⊠ ⊠ x ⊠ x
  x x
  x   x   x
  x x   x
```

<u>Fig. 5.5</u>

Givens' method

First major step gives 7 fill-ins

```
x x     x
o ⊠ x   x
o ⊠   x ⊠ x
  x x
  x   x   x
  x x   x
```

<u>Fig. 5.6</u>

Gaussian elimination

First major step gives 3 fill-ins

5.6 Pivotal strategy

As seen in section 5.5 the Gaussian elimination will normally give less fill-in than the orthogonal methods and among these Givens should be preferred to Householder (see also theoretical results by Duff & Reid ([22]) and Elfving ([32])).

We shall now discuss a pivotal strategy to be used with Givens' method in order to keep the amount of fill-in as small as possible. The strategy is based on an idea by Gentleman ([40]) and has been implemented in the code LLSS01 ([95], [101], [102]).

Assume that we are about to carry out the k-th major step $(1 \leq k \leq n)$

(a) Find the column (number s) with the smallest number $(s_k + 1)$ of active elements (i.e. elements with row number larger than or equal to k).

(b) Interchange columns k and s.

(c) For $i = 1(1)s_k$ find the two rows (with non-zero elements in the pivotal column and) which contain the smallest number of active elements. Create a zero element in one of them using (5.3) - (5.6).

(d) Let row r be the only row which contains a non-zero element in the pivotal column. Interchange rows k and r.

We note that we use a fixed pivotal column and perform the column interchanges before the computations in each major step just like with Gaussian elimination. The row interchanges are performed after the computations in a major step because we vary the pivotal row from one minor step to the other. This is done in order to preserve the sparsity better and to minimize the amount of computation. Note that in contrast to Gaussian elimination we have fill-in in the pivotal row and this would spread if the same row were used again and again.

Remark 5.30 Duff ([17]) has suggested a strategy based on a fixed pivotal row where (c) and (d) are replaced by

(c*) Among the rows with non-zero elements in the pivotal column let row r have a minimal number of non-zero elements. Interchange rows k and r.

(d*) Create zero elements below the diagonal in the pivotal column by using a fixed pivotal row (k) and the other rows in order of increasing number of non-zero elements. ∎

Remark 5.31 Note that the stability criterion (5.3) cannot be used because we have determined beforehand where the zero should be created. ∎

An illustration of the performance of the two strategies with respect to preserving sparsity is given in Fig. 5.7 - 5.11 (using a small matrix and two major steps). The rows that take part in the computations with variable pivotal rows are (1,8), (12, 13), (1,12) in the first major step and (2,9), (2,8), (2,12) in the second major step. Note that in the second step we have not used the option of varying the pivotal row. Still we have fewer fill-ins and easier computations than with the fixed pivotal row.

If the number of non-zero elements in the pivotal column is 3 or less then the number of fill-ins will be the same with the two strategies. Otherwise the variable pivotal row strategy will probably give better results in the sense that the computations leading to the matrix S will be easier, in particular when the matrix is not too sparse, although the number of elements in matrix S may not be much smaller (cf. [17]). Some results in this direction are obtained in [95]. The above statement is illustrated by many numerical experiments in [95].

One drawback with the variable pivotal row strategy is that it cannot be used efficiently when a sequence of coefficient matrices of the same structure are to be factorized. Too much space would be needed in order to keep information about which rows that participate in each minor step.

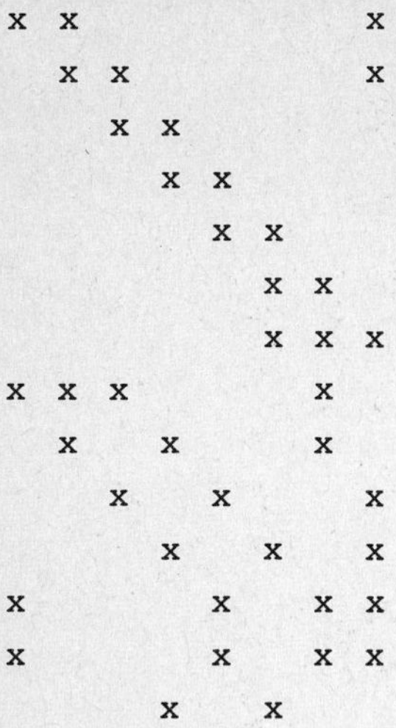

Fig. 5.7

The original matrix

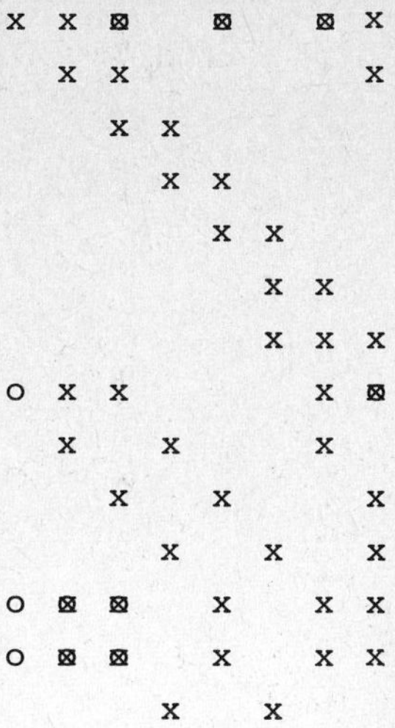

Fig. 5.8

Fixed pivotal row
First major step
3 minor steps
8 fill-ins

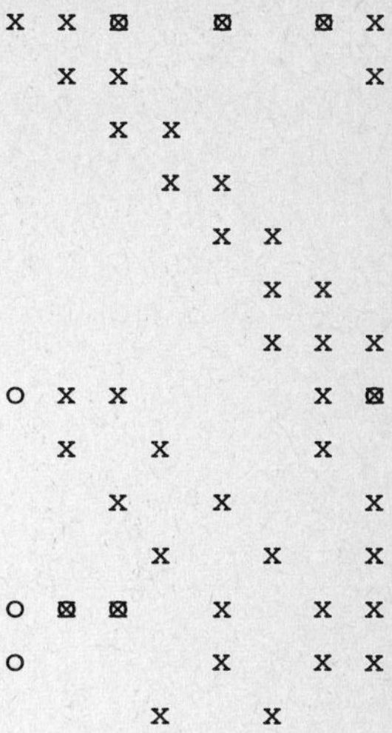

Fig. 5.9

Variable pivotal row
First major step
3 minor steps
6 fill-ins

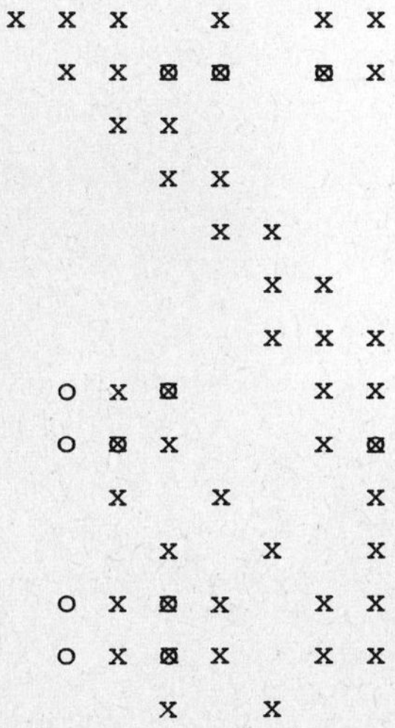

Fig. 5.10

Fixed pivotal row
Second major step
4 minor steps
8 new fill-ins

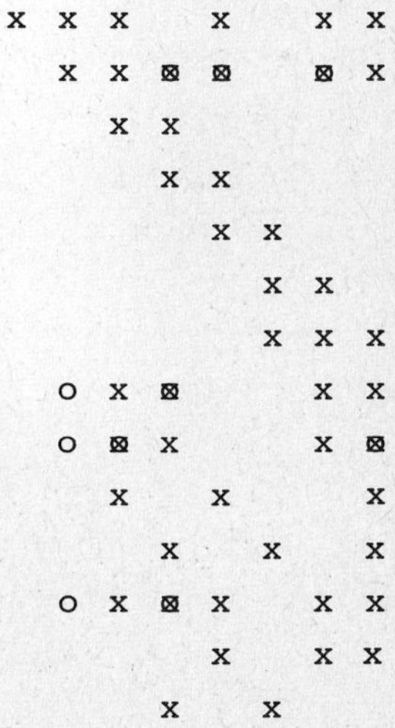

Fig. 5.11

Variable pivotal row
Second major step
3 minor steps
7 new fill-ins

5.7 A 2-stage method based on orthogonal transformations

We shall briefly describe an implementation (the code LLSS01) of the 2-stage method of Example 5.23. In the expression for $y_1 = Hc$ (3.15) the matrix S appears twice and this might be unfavourable since a possible ill-condition of A will be reflected in S. Therefore we combine our method with the method of Example 5.22 from which

$$(7.1) \qquad y_1 = Hb; \quad H = Q_1 S^{-1} D^{-1} R^T P_1.$$

As mentioned earlier we do not want to store the matrix R since it is rather large and probably not very sparse but we can still use (7.1) to compute the direct solution if we perform parallel computations on the right hand side to produce the vector $b^* = R^T P_1 b$ together with the decomposition step.

If we are solving several problems with the same coefficient matrix, one after the other, in particular if we use generalized iterative refinement (4.1) - (4.3) then we use the matrix

$$(7.2) \qquad H = QS^{-1} D^{-2} (S^T)^{-1} Q^T$$

for the succeeding computations. Note that despite (3.14a) the matrix $B_1 = A^T A$ is never calculated. To compute the residuals (4.1) we use

$$(7.3) \qquad r_i^* = b - Ay_i, \qquad r_i = A^T r_i^*.$$

The Gentleman-Givens method is used for the orthogonal decomposition and in order to preserve sparsity better, a drop tolerance, T, is used. The pivotal strategy is based on variable pivotal rows as described in section 5.6.

The storage is arranged in a similar way as described for Gaussian elimination. The arrays A, CNR, RNR, A1, CN are used - the last two to hold the original matrix A. Extra complications arise because we use variable pivotal rows and because fill-ins appear in the pivotal rows. The array HA is split in two parts HA1(m,4) and HA2(n,4) with a total number of columns smaller than the number of columns in HA (13). This is so because we do not store the permutation matrix P_1 nor information about the rows and columns after increasing number of elements, and the pivotal interchanges are organized in a different way.

The inner products in (2.10) and (4.1) - (4.3) are accumulated and stored in double precision as suggested by Björck ([6]). For more details we refer to Zlatev & Nielsen ([101], [102]).

It should be mentioned that two other efficient algorithms for sparse linear least-squares problems have recently been proposed by Björck and Duff [9] and by George and Heath [41]. The last algorithm can easily be modified for the case, where auxiliary storage is to be used; see George, Heath and Plemmons [42].

5.8 Numerical results

The efficiency of the combination: sparse matrix technique + pivotal strategy + large drop tolerance + iterative refinement is demonstrated in several experiments with test matrices of class $F2(m,n,c,r,\alpha)$. In each of the following examples we have kept 4 of the 5 parameters fixed in order to see the effect of varying the last one, and we have used different values of T in order to see the effect it has on storage, time and accuracy.

The experiments were carried out at NEUCC on an IBM 3033 which has $n_1 = 7$ and $n_2 = 16 > 2n_1$ (cf. section 4.6, strategy C). All right-hand sides have been chosen such that the problems were consistent $(r = 0)$ with the solution $x_i = 1$, $i = 1(1)n$. Other right hand sides producing r-vectors with $r_i \in [100,10000]$ have been used with similar results but slightly larger numbers of iterations.

Example 5.32 The matrices are F2(m, 100, 11, 6, 10), m = 100(10)200, so we are changing the ratio m/n. Two values of the drop tolerance have been used (T = 0 and T = 0.01) and from the results in Table 5.12 we see that the larger drop tolerance implies less storage and computing time with roughly the same accuracy. ∎

m	T = 0				T = 0.01			
	COUNT	Time	iter	Accuracy	COUNT	Time	iter	Accuracy
100	3210	1.38	6	5.77E-15	2635	1.06	8	1.93E-14
110	3142	1.42	8	1.78E-14	2619	1.06	8	1.02E-14
120	3801	1.82	7	7.33E-15	2749	1.16	9	1.93E-14
130	3630	1.71	8	2.53E-14	2519	1.01	10	2.15E-14
140	4386	2.32	10	1.82E-14	2460	0.97	10	2.33E-14
150	4326	2.53	9	2.33E-14	2267	0.98	8	7.79E-15
160	4561	2.91	8	1.89E-14	3624	1.93	9	1.31E-14
170	5678	4.75	8	6.89E-14	3623	2.22	9	1.60E-14
180	5675	4.84	6	1.62E-14	4189	2.88	8	1.11E-14
190	5867	5.01	5	3.24E-14	3845	2.57	11	1.75E-14
200	5499	4.97	8	2.35E-14	4323	2.96	9	1.35E-14

Table 5.12

The effect of varying T and m.
A = F2(m, 100, 11, 6, 10); NZ = 6m + 110.

Example 5.33 The matrices are F2(150, 100, c, 6, 10), c = 20(5)65, so we are changing the distribution of the non-zero elements. The results are given in Table 5.13 for T = 0 and T = 0.01. ∎

c	T = 0				T = 0.01			
	COUNT	Time	iter	Accuracy	COUNT	Time	iter	Accuracy
20	5388	3.89	6	1.15E-14	4118	2.40	9	1.38E-14
25	4955	3.62	6	7.78E-15	3975	2.50	8	3.94E-15
30	4933	3.27	7	1.73E-14	3683	2.24	11	2.80E-14
35	5102	3.64	8	2.82E-14	4033	2.57	8	2.93E-14
40	5064	3.74	6	4.20E-14	3441	2.04	9	3.31E-14
45	4002	2.13	9	3.46E-14	2612	1.26	9	3.09E-14
50	3373	1.58	6	1.87E-14	3025	1.57	10	2.93E-14
55	4749	3.11	7	2.95E-14	3478	1.92	10	2.64E-14
60	4672	3.19	11	2.09E-14	3615	2.15	11	2.58E-14
65	4448	2.51	7	1.98E-14	3535	1.95	10	2.38E-14

Table 5.13

The effect of varying T and c.
A = F2(150, 100, c, 6, 10); NZ = 1010.

Example 5.34 The matrices are F2(150, 100, 11, r, 10), r = 5(1)10 so we are changing the width of the band and thus NZ. The results are given in Table 5.14 for T = 0 and T = 0.01. ∎

r	T = 0				T = 0.01			
	COUNT	Time	iter	Accuracy	COUNT	Time	iter	Accuracy
5	4220	2.39	10	2.54E-14	2152	0.75	10	2.84E-14
6	4326	2.53	9	2.33E-14	2267	0.98	8	7.79E-15
7	4717	3.02	7	2.35E-14	3630	2.16	9	2.02E-14
8	4935	3.80	10	2.20E-14	3138	1.60	8	2.13E-14
9	5476	3.96	8	3.37E-14	3331	1.68	9	3.07E-14
10	5889	5.03	7	2.38E-14	3298	1.79	7	2.60E-14

Table 5.14

The effect of varying T and r.
A = F2(150, 100, 11, r, 10); NZ = 150r + 110.

<u>Example 5.35</u> The matrices are F2(125, 100, 11, 5, α), α = 10, 100, 1000, 10000. Since $\max(|a_{ij}|)/\min(|a_{ij}|) = 10\alpha^2$ the matrices with large α are poorly scaled. The accuracy of the solution is given in Table 5.15 for four values of T (0, 0.01, 0.1, 1) and it is seen that large values of T cannot be used with poorly scaled problems. ∎

α	T = 0	T = .01	T = .1	T = 1
10^1	2.02E-14	2.05E-14	3.06E-14	6.82E-7
10^2	4.63E-13	4.45E-13	5.00E- 2	4.72E 0
10^3	2.34E-13	1.02E- 1	3.96E 0	2.32E 2
10^4	3.36E-11	5.51E 0	6.63E 2	3.40E 5

<u>Table 5.15</u>

The effect on the accuracy of varying T and α.
A = F2(125, 100, 11, 5, α); NZ = 735.

<u>Example 5.36</u> Same matrices as in Example 5.35. We compare the direct solution (DS) and iterative refinement (IR) with T = 0 and T = 0.01. The results are given in Table 5.16. ∎

	DS, T = 0			IR, T = 0			IR, T = 0.01		
α	COUNT	Time	Accuracy	COUNT	Time	Accuracy	COUNT	Time	Accuracy
10^1	3308	1.36	8.66E-4	3308	1.53	2.02E-14	1930	0.72	2.05E-14
10^2	3306	1.35	1.20E-3	3306	1.52	4.63E-13	2048	0.91	4.45E-13
10^3	3308	1.36	8.50E-2	3308	1.51	2.34E-13	2101	0.78	1.02E- 1
10^4	3308	1.36	4.42E 0	3308	1.59	3.36E-11	2275	0.89	5.51E 0

<u>Table 5.16</u>

Comparison of DS and IR with different α.
A = F2(125, 100, 11, 5, α); NZ = 735.

We can draw the following general conclusions from the experiments.

(a) When the iterative refinement process is convergent we can expect $2n_1$ digits accuracy (cf. [3], [4] and [6]).

(b) IR plus a large T leads to a reduction in both storage and computation time for the method of section 5.7.

(c) DS is slightly faster than IR with the same T but the accuracy is considerably better with IR.

(d) IR may converge even if the matrix is very poorly scaled but a smaller value of T should be used.

The above conclusions from experiments with the method of section 5.7 are in very good agreement with our results from Chapter 4 relating to Gaussian elimination with square matrices. We can in this connection note that row scaling normally is not allowed with least squares problems, so we shall have to live with poorly scaled matrices (and smaller values of T).

Appendix: The codes used in the text

In this appendix we list the codes which we have used throughout the text, the program libraries where they can be found, and the computing centres where test runs have been performed.

Computing centres:

NEUCC	-	Northern Europe University Computing Centre, Technical University, Lyngby, Denmark.
RECAU	-	Regional Computing Centre, Aarhus University, Denmark.
RECKU	-	Regional Computing Centre, Copenhagen University, Denmark.

Program Libraries:

Harwell Library	-	Developed at AERE, Harwell, England. Implemented at NEUCC.
NAG Library	-	Developed by Numerical Algorithms Group, Oxford, England, [64]. Implemented at RECAU and RECKU.

Codes:

MA18	-	This package solves linear systems with general sparse matrices. The package is described in [15] and is a standard subroutine in the Harwell Library.
MA28	-	This package solves linear systems with general sparse matrices. The package is described in [18] and is a standard subroutine in the Harwell Library.

F04ACE/F	-	This subroutine solves linear systems with symmetric, positive definite band-matrices and is a standard subroutine in the NAG Library (Mark 7). It can be considered as a NAG version of the subroutine described in [88] pp. 50-56.
F01BRE/F + F04AXE/F	-	A set of subroutines to solve linear systems with general sparse matrices. These subroutines are standard subroutines in the NAG Library (Mark 7) and can be considered as NAG versions of MA28.
INDANL + INDOPR	-	A set of subroutines to solve symmetric, indefinite linear systems. The subroutines are described in [62] and can be obtained from the Institute for Numerical Analysis, Technical University, Lyngby, Denmark.
ST	-	This subroutine solves linear systems with general sparse matrices. The subroutine is described in [105] and can be obtained from the Institute for Numerical Analysis, Technical University, Lyngby, Denmark.
SSLEST	-	This package solves linear systems with general sparse matrices and exists in two versions. One is written in ALGOL W, is described in [97] and is a standard subroutine in the ALGOL W Library at NEUCC. The second version is written in FORTRAN and described in [98]. Both versions can be obtained from the Institute for Numerical Analysis, Technical University, Lyngby, Denmark.
SIRSM	-	This package solves linear systems with general sparse matrices by iterative refinement. The package is described in [99], [100] and can be obtained from the Institute for Numerical Analysis, Technical University, Lyngby, Denmark.

LLSS01 - This package solves linear least squares problems by iterative refinement. The package is described in [95], [101], [102] and can be obtained from the Institute for Numerical Analysis, Technical University, Lyngby, Denmark.

Y12M - This package solves linear systems with general sparse matrices directly or by iterative refinement. The package is described in [103], [108]. The subroutines are implemented as standard subroutines at RECKU and can be obtained from RECKU.

<u>Remark 1</u> - All codes except LLSS01 are based on some version of Gaussian elimination. LLSS01 uses the Gentleman-Givens version of plane rotations.

<u>Remark 2</u> - All codes are written in FORTRAN. An ALGOL W version of SSLEST is also available.

<u>Remark 3</u> - MA28 is more efficient than MA18 (see section 2.8 and [25]). Note too that MA28 can perform transformation to block triangular form if this is possible.

<u>Remark 4</u> - Y12M is superior to ST, SSLEST and SIRSM. This is the only package where a drop tolerance and iterative refinement can be used as an option. An experimental version of SSLEST with these options is under development.

<u>Remark 5</u> - Other codes for sparse problems have been described by Duff [19], [20].

References

1. O. Axelsson: A generalized SSOR method, BIT 12 (1972) 443-467.

2. O. Axelsson: On preconditioned conjugate gradient methods. "Conjugate Gradient Methods and Similar Techniques" (I.S. Duff, ed.), A.E.R.E. R9636, Harwell, England (1979) 23-35.

3. Å. Björck: Iterative refinement of linear least squares solutions I, BIT 7 (1967) 257-278.

4. Å. Björck: Iterative refinement of linear least squares solutions II, BIT 8 (1968) 8-30.

5. Å. Björck: Methods for sparse linear least squares problems. "Sparse Matrix Computations" (J. Bunch & D. Rose, eds.) Academic Press, New York (1976) 177-199.

6. Å. Björck: Comment on the iterative refinement of least-squares solutions, J. Amer. Statist. Assoc. 73 (1978) 161-166.

7. Å. Björck: Numerical algorithms for linear least squares problems, Report 2, Matematisk Institut, Univeritetet i Trondheim, Trondheim, Norway (1978).

8. Å. Björck & T. Elfving: Accelerated projection methods for computing pseudoinverse solutions of systems of linear equations, BIT 19 (1979) 145-163.

9. Å. Björck & I.S. Duff: A direct method for the solution of sparse linear least squares problems, Lin. Alg. Appl. 34 (1980) 43-67.

10. Å. Björck & G.H. Golub: ALGOL programming, contribution No. 22: Iterative refinement of linear least square solutions by Householder transformation, BIT 7 (1967) 322-337.

11. R.K. Brayton, F.G. Gustavson & R.A. Willoughby: Some results on sparse matrices, Math. Comp. 24 (1970) 937-954.

12. R.J. Clasen: Techniques for automatic tolerance control in linear programming, Comm. ACM 9 (1966) 802-803.

13. A.K. Cline, A.R. Conn & C.F. Van Loan: Generalizing the LINPACK condition estimator. "Numerical Analysis", (J.P. Hennart, ed.), Lecture Notes in Mathematics 909, Springer, Berlin (1982) 73-83.

14. A. K. Cline, C.B. Moler, G.W. Stewart & J.H. Wilkinson: An estimate for the condition number of a matrix, SIAM J. Numer. Anal. 16 (1979) 368-375.

15. A.R. Curtis & J.K. Reid: The solution of large sparse unsymmetric systems of linear equations, J. Inst. Math. Appl. 8 (1971) 344-353.

16. J.J. Dongarra, J.R. Bunch, C.B. Moler & G.W. Stewart: LINPACK - Users' Guide, SIAM, Philadelphia (1979).

17. I.S. Duff: Pivot selection and row ordering in Givens reduction on sparse matrices, Computing 13 (1974) 239-248.

18. I.S. Duff: MA28 - a set of FORTRAN subroutines for sparse unsymmetric linear equations, A.E.R.E. R8730, Harwell, England (1977).

19. I.S. Duff: Practical comparisons of codes for the solution of sparse linear systems. "Sparse Matrix Proceedings 1978" (I.S. Duff & G.W. Stewart, eds.) SIAM, Philadelphia (1979) 107-134.

20. I.S. Duff: A survey of sparse matrix software. "Sources and Development of Mathematical Software" (W.R. Cowell, ed.), Prentice-Hall, Englewood Cliffs, N.J. (to appear).

21. I.S. Duff & J.K. Reid: A comparison of sparsity orderings for obtaining a pivotal sequence in Gaussian elimination, J. Inst. Math. Appl. 14 (1974) 281-291.

22. I.S. Duff & J.K. Reid: On the reduction of sparse matrices to condensed forms by similarity transformations, J. Inst. Math. Appl. 15 (1975) 217-224.

23. I.S. Duff & J.K. Reid: A comparison of some methods for the solution of sparse overdetermined systems of linear equations, J. Inst. Math. Appl. 17 (1976) 267-280.

24. I.S. Duff & J.K. Reid: Performance evaluation of codes for sparse matrix problems. "Performance Evaluation of Numerical Software" (L.D. Fosdick, ed.) IFIP, North-Holland (1979) 121-135.

25. I.S. Duff & J.K. Reid: Some design features of a sparse matrix code, ACM Trans. Math. Software 5 (1979) 18-35.

26. I.S. Duff & J.K. Reid: The multifrontal solution of indefinite sparse symmetric linear systems, A.E.R.E. CSS 122, Harwell, England (1982).

27. S.C. Eisenstat: Efficient implementation of a class of preconditioned conjugate gradient methods, SIAM J. Sci. Statist. Comput. 2 (1981) 1-4.

28. S.C. Eisenstat, M.C. Gursky, M.H. Schultz & A.H. Sherman: The Yale sparse matrix package I. The symmetric codes, Report 112, Department of Computer Science, Yale University (1977).

29. S.C. Eisenstat, M.C. Gursky, M.H. Schultz & A.H. Sherman: The Yale sparse matrix package II. The non-symmetric codes, Report 114, Department of Computer Science, Yale University (1977).

30. S.C. Eisenstat, M.H. Schultz & A.H. Sherman: Algorithms and data structures for sparse symmetric Gaussian elimination, SIAM J. Sci. Statist. Comput. 2 (1981) 225-237.

31. S.C. Eisenstat, M.H. Schultz & A.H. Sherman: Software for sparse Gaussian elimination with limited core storage. "Sparse Matrix Proceedings 1978" (I.S. Duff & G.W. Stewart, eds.), SIAM, Philadelphia (1979) 135-153.

32. T. Elfving: A note on sparsity in Gauss and Givens methods, LITH-MAT-R-1976-5, Department of Mathematics, Linköping University, Sweden (1976).

33. D.J. Evans: The use of preconditioning in iterative methods for solving linear equations with symmetric positive definite matrices, J. Inst. Math. Appl. 4 (1968) 295-314.

34. D.J. Evans: Iterative sparse matrix algorithms. "Software for Numerical Mathematics" (D.J. Evans, ed.), Academic Press, London (1974) 49-83.

35. G.E. Forsythe, M.A. Malcolm & C.B. Moler: Computer Methods for Mathematical Computations, Prentice-Hall, Englewood Cliffs, N.J. (1977).

36. G.E. Forsythe & C.B. Moler: Computer Solution of Linear Algebraic Systems, Prentice-Hall, Englewood Cliffs, N.J. (1967).

37. C.W. Gear: Numerical error in sparse linear equations, UIUCDCS-F-75-885, Department of Computer Science, University of Illinois at Urbana-Champaign (1975).

38. W.M. Gentleman: Least squares computations by Givens transformations without square roots, J. Inst. Math. Appl. 12 (1973) 329-336.

39. W.M. Gentleman: Error analysis of QR decompositions by Givens transformations, Lin. Alg. Appl. 10 (1975) 189-197.

40. W.M. Gentleman: Row elimination for solving sparse linear systems and least squares problems. "Numerical Analysis Dundee 1975" (G.A. Watson, ed.), Lecture Notes in Mathematics 506, Springer, Berlin (1976) 122-133.

41. J.A. George & M.T. Heath: Solution of sparse linear least squares problems using Givens rotations, Lin. Alg. Appl. 34 (1980) 69-83.

42. J.A. George, M.T. Heath & R.J. Plemmons: Solution of large - scale sparse linear least squares problems using auxiliary storage, SIAM J. Sci. Statist. Comput. 2 (1981) 416-429.

43. J.A. George & J.W. Liu: Computer Solution of Large Sparse Positive Definite Systems, Prentice-Hall, Englewood Cliffs, N.J. (1981).

44. J.A. George, J.W. Liu & E. Ng: User guide for SPARSPAK: Waterloo sparse linear equations package, CS - 78 - 30, Department of Computer Science, University of Waterloo, Canada (1980).

45. J.W. Givens: Numerical computation of the characteristic values of a real symmetric matrix, ORNL-1574 Oak Ridge National Laboratory (1954).

46. J.W. Givens: Computation of plane unitary rotations transforming a general matrix to triangular form, J. Soc. Ind. Appl. Math. 6 (1958) 26-50.

47. G.H. Golub & W. Kahan: Calculating the singular values and pseudo-inverse of a matrix, J. SIAM Ser. B, Numer. Anal. 2 (1965) 205-224.

48. G.H. Golub & C. Reinsch: Singular value decomposition and least squares solutions, Numer. Math. 14 (1970) 403-420.

49. G.H. Golub & J.H. Wilkinson: Note on the iterative refinement of least squares solution. Numer. Math. 9 (1966) 139-148.

50. R.G. Grimes & J.G. Lewis: Condition number estimation for sparse matrices, SIAM J. Sci. Statist. Comput. 2 (1981) 384-388.

51. F.G. Gustavson: Some basic techniques for solving sparse systems of linear equations. "Sparse Matrices and Their Applications" (D.J. Rose & R.A. Willoughby, eds.), Plenum Press, New York (1972) 41-52.

52. F.G. Gustavson: Two fast algorithms for sparse matrices: multiplication and permuted transposition, ACM Trans. Math. Software 4 (1978) 250-269.

53. S. Hammarling: A note on modifications to the Givens plane rotation, J. Inst. Math. Appl. 13 (1974) 215-218.

54. A.S. Householder: Unitary triangularization of a nonsymmetric matrix, J. Assoc. Comput. Mach. 5 (1958) 339-342.

55. M. Jankowski & H. Woźniakowski: Iterative refinement implies numerical stability, BIT 17 (1977) 303-311.

56. A. Jennings & G.M. Malik: Partial elimination, J. Inst. Math. Appl. 20 (1977) 307-316.

57. H.M. Markowitz: The elimination form of the inverse and its application to linear programming, Management Sci. 3 (1957) 255-269.

58. J.A. Meijerink & H.A. van der Vorst: An iterative solution method for linear systems of which the coefficient matrix is a symmetric M-matrix, Math. Comp. 31 (1977) 148-162.

59. C.B. Moler: Iterative refinement in floating point, J. Assoc. Comput. Mach. 14 (1967) 316-321.

60. C.B. Moler: Three research problems in numerical linear algebra. "Numerical Analysis" (G.H. Golub & J. Oliger, eds.), American Mathematical Society, Providence, Rhode Island (1978) 1-18.

61. E.H. Moore: On the reciprocal of the general algebraic matrix, Bull. Amer. Math. Soc. 26 (1920) 394-395.

62. N. Munksgaard: Fortran subroutines for direct solution of sets of sparse and symmetric linear equations, NI-77-05, Institute for Numerical Analysis, Technical University of Denmark, Lyngby, Denmark (1977).

63. N. Munksgaard: Solving sparse symmetric sets of linear equations by preconditioned conjugate gradients, ACM Trans. Math. Software 6 (1980) 206-219.

64. NAG Library Fortran Manual, Mark 7, vol. 3-4, Numerical Algorithms Group, Banbury Road, Oxford, England (1979).

65. R. Penrose: A generalized inverse for matrices, Proc. Cambridge Phil. Soc. 51 (1955) 406-413.

66. G. Peters & J.H. Wilkinson: The least squares problem and pseudo-inverses, Computer J. 13 (1970) 309-316.

67. J.K. Reid: A note on the stability of Gaussian elimination, J. Inst. Math. Appl. 8 (1971) 374-375.

68. J.K. Reid: Fortran subroutines for handling sparse linear programming bases, A.E.R.E. R8269, Harwell, England (1976).

69. J.K. Reid: Solution of linear systems of equations: direct methods (general). "Sparse Matrix Techniques" (V.A. Barker, ed.), Lecture Notes in Mathematics 572, Springer, Berlin (1977) 102-129.

70. K. Schaumburg & J. Wasniewski: Use of a semiexplicit Runge-Kutta integration algorithm in a spectroscopic problem, Computers and Chemistry 2 (1978) 19-24.

71. K. Schaumburg, J. Wasniewski & Z. Zlatev: Solution of ordinary differential equations with time dependent coefficients. Development of a semiexplicit Runge-Kutta algorithm and application to a spectroscopic problem, Computers and Chemistry 3 (1979) 57-63.

72. K. Schaumburg, J. Wasniewski & Z. Zlatev: The use of sparse matrix technique in the numerical integration of stiff systems of linear ordinary differential equations, Computers and Chemistry 4 (1980) 1-12.

73. A.H. Sherman: Algorithms for sparse Gaussian elimination with partial pivoting, ACM Trans. Math. Software 4 (1978) 330-338.

74. R.D. Skeel: Scaling for numerical stability in Gaussian elimination, J. Assoc. Comput. Mach. 26 (1979) 494-526.

75. R.D. Skeel: Iterative refinement implies numerical stability for Gaussian elimination, Math. Comp. 35 (1980) 817-832.

76. R.D. Skeel: Effect of equilibration on residual size for partial pivoting, SIAM J. Numer. Anal. 18 (1981) 449-454.

77. G.W. Stewart: Introduction to Matrix Computations, Academic Press, New York (1973).

78. G.W. Stewart: The economical storage of plane rotations, Numer. Math. 25 (1976) 137-138.

79. G.W. Stewart: On the perturbation of pseudo-inverses, projections, and linear least squares problems, SIAM Review 19 (1977) 634-662.

80. R.P. Tewarson: Sparse Matrices, Academic Press, New York (1973).

81. A.D. Tuff & A. Jennings: An iterative method for large systems of linear structural equations, Internat. J. Numer. Methods Engrg. 7 (1973) 175-183.

82. J. Wasniewski, Z. Zlatev & K. Schaumburg: A method for reduction of the storage requirement by the use of some special computer facilities; application to linear systems of algebraic equations, Computers and Chemistry 6 (1982) 181-192.

83. V.V. Voevodin: Computational Bases of the Linear Algebra, Nauka, Moscow (in Russian) (1977).

84. J.H. Wilkinson: Error analysis of direct methods of matrix inversion, J. Assoc. Comput. Mach. 8 (1961) 281-330.

85. J.H. Wilkinson: Rounding Errors in Algebraic Processes, Prentice-Hall, Englewood Cliffs, N.J. (1963).

86. J.H. Wilkinson: The Algebraic Eigenvalue Problem, Oxford University Press, London (1965).

87. J.H. Wilkinson: Some recent advances in numerical linear algebra. "The State of the Art in Numerical Analysis" (D.A.H. Jacobs, ed.) Academic Press, New York (1977) 3-53.

88. J.H. Wilkinson & C. Reinsch: Handbook for Automatic Computation, vol. 2, Linear Algebra, Springer, Berlin (1971).

89. P. Wolfe: Error in the solution of linear programming problems. "Error in Digital Computation" (L.B. Rall, ed.), Vol. 2, Wiley, New York (1965) 271-284.

90. D.M. Young: Iterative Solution of Large Linear Systems, Academic Press, New York (1971).

91. D.M. Young & D.R. Kincaid: The ITPACK package for large sparse linear systems. "Elliptic Problem Solvers" (M. Schultz, ed.) Academic Press, New York (1981) 163-185.

92. Z. Zlatev: On some pivotal strategies in Gaussian elimination by sparse technique, SIAM J. Numer. Anal. 17 (1980) 18-30.

93. Z. Zlatev: On solving some large linear problems by direct methods, DAIMI PB-111, Department of Computer Science, University of Aarhus, Denmark (1980).

94. Z. Zlatev: Modified diagonally implicit Runge-Kutta methods, SIAM J. Sci. Statist. Comput. 2 (1981) 321-334.

95. Z. Zlatev: Comparison of two pivotal strategies in sparse plane rotations, Comput. Math. Appl. 8 (1982) 119-135.

96. Z. Zlatev: Use of iterative refinement in the solution of sparse linear systems, SIAM J. Numer. Anal. 19 (1982) 381-399.

97. Z. Zlatev & V.A. Barker: Logical procedure SSLEST - an Algol W procedure for solving sparse systems of linear equations, NI-76-13, Institute for Numerical Analysis, Technical University of Denmark, Lyngby, Denmark (1976).

98. Z. Zlatev, V.A. Barker & P.G. Thomsen: SSLEST: A FORTRAN IV subroutine for solving sparse systems of linear equations (User's guide). NI-78-01, Institute for Numerical Analysis, Technical University of Denmark, Lyngby, Denmark (1978).

99. Z. Zlatev & H.B. Nielsen: Preservation of sparsity in connection with iterative refinement, NI-77-12, Institute for Numerical Analysis, Technical University of Denmark, Lyngby, Denmark (1977).

100. Z. Zlatev & H.B. Nielsen: SIRSM - a package for the solution of sparse systems by iterative refinement, NI-77-13, Institute for Numerical Analysis, Technical University of Denmark, Lyngby, Denmark (1977).

101. Z. Zlatev & H.B. Nielsen: LLSS01 - a FORTRAN subroutine for solving linear least squares problems (User's guide), NI-79-07, Institute for Numerical Analysis, Technical University of Denmark, Lyngby, Denmark (1979).

102. Z. Zlatev & H.B. Nielsen: Least squares solution of large linear problems. "Symposium i Anvendt Statistik 1980" (A. Höskuldsson et. al. eds.), NEUCC, Lyngby, Denmark (1980) 17-52.

103. Z. Zlatev, K. Schaumburg & J. Wasniewski: Implementation of an iterative refinement option in a code for large and sparse systems, Computers and Chemistry 4 (1980) 87-99.

104. Z. Zlatev, K. Schaumburg and J. Wasniewski: A testing scheme for subroutines solving large linear problems, Computers and Chemistry 5 (1981) 91-100.

105. Z. Zlatev & P.G. Thomsen: ST - a Fortran IV subroutine for the solution of large systems of linear algebraic equations with real coefficients by use of sparse technique, NI-76-05, Institute for Numerical Analysis, Technical University of Denmark, Lyngby, Denmark (1976).

106. Z. Zlatev & P.G. Thomsen: Application of backward differentiation methods to the finite element solution of time-dependent problems, Internat. J. Numer. Methods Engrg. 14 (1979) 1051-1061.

107. Z. Zlatev & P.G. Thomsen: Sparse matrices - efficient decomposition and applications. "Sparse Matrices and Their Uses" (I.S. Duff, ed.) Academic Press, London (1981) 367-375.

108. Z. Zlatev, J. Wasniewski & K. Schaumburg: Y12M - Solution of Large and Sparse Systems of Linear Algebraic Equations, Lecture Notes in Computer Science 121, Springer, Berlin (1981).

109. Z. Zlatev, J. Wasniewski & K. Schaumburg: Comparison of two algorithms for solving large linear systems, SIAM J. Sci. Statist. Comput. 3 (1982) 486-501.

110. D.J. Evans: The analysis and application of sparse matrix algorithms in the finite element method. "The Mathematics of Finite Elements and Applications" (J.R. Whiteman, ed.) Academic Press, London (1973) 427-447.

111. H.B. Nielsen: Iterative refinement, NI-76-02, Institute for Numerical Analysis, Technical University of Denmark, Lyngby, Denmark (1976).

L.-Brault

DATE DE RETOUR

20 FEV. 1989			
28 OCT. 1994			
16 SEP. 1997			
07 NOV 1997			
27 MARS 1998			

[redacted]